GLACIERS OF NORTH AMERICA

Climbers ascend Mt. Baker in Washington state. Photo by Asahel Curtis, June 20, 1908. Courtesy of the Washington State Historical Society, Tacoma.

GLACIERS
OF
NORTH AMERICA
A FIELD GUIDE

SUE A. FERGUSON

Fulcrum Publishing
Golden, Colorado

Cover and interior design by Patty Maher

Cover photograph: The Marjorie Glacier, Glacier Bay, Alaska
© 1992 David B. Palmer

All illustrations by Sue A. Ferguson

Library of Congress Cataloging-in-Publication Data
Ferguson, Sue A.
Glaciers of North America : a field guide / Sue A. Ferguson.
p. cm.
Includes bibliographical references and index.
ISBN 1-55591-075-0 (pbk.)
1. Glaciers—North America. I. Title.
GB2412.F47 1992
551.3'12'097—dc20 91–58484
CIP

Printed in the United States of America
0 9 8 7 6 5 4 3 2 1

Fulcrum Publishing
350 Indiana Street, Suite 350
Golden, Colorado 80401

To Pop,
who helped show me the way to adventure

Table of Contents

Foreword

Edward R. LaChapelle

Reading *Glaciers of North America* set me to thinking about my earliest encounters with glaciers, some fifty years back. A lot of glaciers have flowed down the mountains since, but the earliest ones still hover like ghosts in the landscape of my mind.

Early in the summer of 1942, I started working at Paradise Lodge on the flanks of Mt. Rainier. The first few days I found the weather far from that of the promised paradise: foggy and rainy, with a constant cold wind. One evening the clouds finally broke and the snow and ice of Mt. Rainier's glaciers glittered in the setting sun. I was spellbound. A whole new world gripped my psyche and has never let go. I had to explore it. Every chance I got was spent hiking and climbing on and around the glaciers. I soon learned that smooth-soled shoes and light clothing were definitely not glacier wear. Together with a couple of friends, I set about improvising gear that today's well-equipped mountaineer would view with horror. For boots I used my father's old logging boots with caulked soles. We heard they used ice axes in Switzerland; we settled for *alpenstocks* (broom handles with spikes in one end). The best rope we could come up with was half-inch sisal, heavy when dry, like lead when wet, an iron bar when frozen. As to the way this equipment was to be used, we were on our own, learning by trial and many errors. Youthful resilience and bountiful measure of good luck brought us through more or less intact and allowed us to gain an understanding of the ever-changing character of snow and ice.

What I didn't have then was the company of people like Sue Ferguson or even a field guide, much less a guide as instructive as this one. I marvel at what my life in those days (plus my future career in glaciology) would have been like with such practical and useful guidance to set me on my way.

My career in glaciology paralleled the emergence of the field as a recognized subdiscipline of science after World War II. At one time a subject of scattered interest on the part of geologists and geographers, the full interdisciplinary nature of glacier studies came to fruition during the International Geophysical Year of 1957. Meteorology, climatology, hydrology, mechanics and thermodynamics all played a role in understanding the behavior of glaciers and their responses to climate change. Glaciology become respectable as an independent profession, usually entered through geology, geophysics, physics, or atmospheric science. A generation of scientists has evolved during the postwar period that carries on a diverse range of teaching, research, and practical applications of the growing knowledge about snow and ice. This field guide reflects this breadth of knowledge, bringing to the reader in layperson's terms an informed understanding of how glaciers work, where they are, and how to travel on them. Sue Ferguson is one of those modern glacier scientists who brings wide experience and an impeccable background in geophysics to the task of revealing the truths about North American glaciers. Read this guide with confidence.

All these modern scientific advances haven't entirely eclipsed the practical knowledge gained from our earlier and simpler relation to glaciers. I would like to share an arcane piece of practical knowledge with the handsomely equipped modern mountaineer and glacier traveller. If you want to get optimum traction for walking on ice without the cumbersome use of crampons, wear a pair of logging boots with caulked soles.

Acknowledgments

In gathering material for this book I had the fortune to become reacquainted with those I had met early in my career and to meet new sorcerers of the névé névé land. I am grateful to all who took the time to answer silly questions, discuss controversial topics, and offer much-needed information. I expressly want to thank Dr. Charles Raymond and Simon Ommanney for their invaluable suggestions. Also, Nick Parker, Jos Lang, Colin Monteath, Tad Pfeffer, and Howard Conway set me straight on a few concepts of safe travel. Special thanks to Bob Krimmel for giving me an easy ride onto the South Cascade Glacier and for leading the tangled, overgrown route out.

Without initial encouragement from Betsy Armstrong and her continued friendship, I would never have started this project. So thanks, Betsy, I think. Rich Metcalf taught me the first rule of glacier etiquette, "Never complain." That rule has helped me through some very sticky situations. In addition, Rich gave me my first tattered copy of Paterson's *The Physics of Glaciers*. Even though a few pages are missing, it has been my "bible" for many years.

Ellen Grant provided helpful perspective and needed support throughout this project. She also came through with scrumptious pâté to help me over the critical times.

Finally, and with undying gratitude, I must thank Dr. Edward LaChapelle. He suffered through the first draft of the manuscript. The fervor of his cartoon critique prompted me to replenish his supply of red ink. Ed has an uncanny way of supporting my wild ideas and at the same time helping me to find and destroy factoids.

Conversion Factors

Altitude

meters X 3.281 = feet

Area

square kilometers X 0.386 = square miles
square kilometers X 247.105 = acres

Length

millimeters X 0.0394 = inches
centimeters X 0.394 = inches
meters X 3.281 = feet
kilometers X 0.622 = miles

Speed

meters per second X 2.237 = miles per hour

Temperature

(degrees Centigrade X 9/5) + 32 = degrees Fahrenheit

Volume

liters X 1.057 = liquid quarts
liters X 0.264 = liquid gallons

Weight

kilograms X 2.205 = pounds
milligrams X 3.527 x 10^{-5} = ounces

Density

kilograms per cubic meter X 0.0624 = pounds per cubic foot

Note: English unit conversions in the text are approximated to the nearest significant digit.

CHAPTER
—1—

INTRODUCTION

There was no sound and the air was as white as the snow. All I could see was the faint, fog-filtered red glow of my nearby tent. Then a bird gently landed on the tips of my skis. She stayed only long enough to catch her breath. In a whisper, as if not meaning to intrude, her feathers ruffled in preparation to leave, back into the surrounding cloud. Her flapping gray wings pushed away the droplets of fog, and I saw a wake form behind her as she gradually disappeared.

The quiet returned—such a contrast to the blizzard that had raged all night. With no distinction between the dense fog and the knee-deep fresh snow, I felt weightless in an ethereal space. And here I was, floating on top of a mass of ice that was as tall as the Eiffel Tower.

I shuffled through the snow toward the edge of our camp and peered into the whiteness, just making out the deep shadow of an ominous crevasse. While I was trying to imagine the depth and breadth of this cavern, the rest of my team began to crawl out of their tents. The silence was broken as we began preparations to finish our trek to base camp on this, my first glacier research expedition.

Growing up in the Pacific Northwest, I have always enjoyed glaciers—exploring the Paradise ice caves, touring the Columbia Icefield by Sno-Cat™, glissading down the Challenger Glacier after a summer hike, taking a cool drink from the dripping snout of the Hume Glacier—but my scientific curiosity about glaciers did not blossom until I participated on some research projects in graduate school. With my education in physics and snow and ice mechanics, the depth of my enjoyment and wonderment grew. I became captivated.

Even now, from my home in Seattle, when I look out the window at the glaciers on Mt. Rainier, my heart flutters with awe. Millions of tons of ice creep and grind down that mountain every year. During the summer its glaciers make Rainier look like Medusa's head with their dirt-covered tongues snaking into the forested slopes below. Fortunately, a quick coat of fresh winter snow renews its pristine whiteness every season.

Up close, the grandeur of a glacier is amplified. Large ice crystals with strange and unusual shapes can be found in nooks and crannies. Huge ice cliffs and deep blue-walled crevasses are common. Striped and swirling formations of crushed rock display beautiful patterns.

This is not necessarily a quiet display of grandeur. While many glaciers creep silently along, others crack and groan so much as they grind downhill it's hard not to be frightened. Even if a glacier is relatively quiet, thundering avalanches may crash through the silence.

Feeling glaciers is as inspiring as seeing or hearing them. If you have ever ridden a bicycle in Alaska around Turnagin Arm from Girdwood to Seward, you will know what I mean. Upon turning the corner at Portage, you are blasted with a rush of cold air draining off the Portage Glacier. The glacier is several miles away but its katabatic breath (caused by air cooling over the ice, becoming heavy, and draining down the valley) is no less refreshing.

Yet there are no smells. With only a little algae and some small worms to provide a bouquet, glaciers are one of the cleanest smelling places on earth.

I'd like to share my love for glaciers with you. This book is about glaciers—what they are, where to find them, what to look for, and how to stay out of trouble.

Once the powerful natural processes that create glacier phenomena are understood, the elegance of even the smallest glacier is further heightened. For this reason the first part of *Glaciers of North America* is devoted to explaining the origin and natural development of a variety of glacial phenomena. Basic physical processes are described in lay terms and accompanied by real-life examples from our many North American glaciers. Also, simple at-home and on-ice experiments are included to complement explanations.

The following chapters are organized to help unfold the mystery of glaciers for you as a field observer. First we'll define several varieties of glaciers that can be found in North America and help you decide whether that little snow patch you enjoy every summer is a glacier or not. We will discover how a glacier can form from the accumulation of billions of tiny little snowflakes and how it can die, old and crusty. Next we'll survey the size and extent of glaciers around North America from a little no-name in Montana to the great Malaspina in Alaska. We'll check out the ages of these patches of ice, what makes them advance and retreat, and how you can learn of past climates from a glacier's ice layers.

After this introductory information, we'll dive into the nitty gritty about how glaciers move and what causes them to surge, crack, and fall apart. Water on and within a glacier plays a major part in sculpting its many features and we'll investigate common moulins and unusual jökulhlaups and geysers. Other features, like sun cups and moraines, will also be explained.

To help you look at glaciers safely I have also included a chapter on expected dangers and recommended methods to help stay out of trouble. No book can adequately cover all the practices and techniques needed to follow a safe route through the mountains. However, I hope that this chapter provides enough information to show first-timers the scope of potential hazards for which they must be prepared. In addition, these words may offer an opportunity for the experienced to reaffirm his or her dedication to safety.

There are several hundred thousand mountain glaciers in North America. Many have a secret lore and mysticism. Some provide water for cities and towns. Others can impact shipping by releasing icebergs into our near-shore waters. Still others offer year-round recreational opportunities.

The last section of *Glaciers of North America* includes an inventory of accessible glaciers. Although there are only a few dozen glaciers that can be reached easily, there is enough variety to interest the scientific minded, fascinate the artist, challenge the explorer, and amaze the casual observer. The location and method of accessing each glacier is described. Its main features are summarized and any interesting anecdotes are included.

Finally, a glossary is included to help you quickly locate the meaning of a strange word. The cross-referenced index should help you locate additional information on features that you see in the field. And a selected bibliography offers a list of other interesting books and articles on glaciers and related subjects.

I hope you find glaciers as fun and fascinating as I do.

CHAPTER
—2—

Real Glaciers Eat Rocks

I used to give glacier talks on Sunday afternoons to visitors of Olympic National Park. I would walk down from the research hut where I was stationed for the summer, to a rocky ledge overlooking the Blue Glacier. There I would wait for a group of hikers to come up from Glacier Meadows, a campsite just a few minutes away.

My vantage point offered visitors their first complete view of the glacier. Before anybody had a chance to ask, "What is a glacier?" I pointed to the spectacular vista of icefalls, crevasses, and the long icy tongue reaching far down the valley and said, "This is a glacier." Everybody knew what I meant.

Recently, while visiting the St. Mary's Glacier in Colorado, I was asked the same question, "What is a glacier?" This time I had trouble answering. This little summer ski field had few of the features I have come to expect in a glacier.

Although I would like to itemize the exact characteristics required for the St. Mary's to qualify as a real glacier, it is not that easy. Even scientists cannot agree on a precise definition for glaciers. Be that as it may, I suspect you want to know what a glacier is, so let us try to sort out some of the terminology.

A Glacier Is a Glacier Is a Glacier

The one thing all seem to agree on is that a glacier is a large mass of ice that originates on land. *Ice* is the key word here. A perennial patch of snow (that is, one lasting from year to year) just will not cut it.

Although the term *large* is not completely defined, at least one international group stipulates that a glacier must have a surface area larger than one-tenth of a square kilometer (25 acres). If it were roundish, a glacier this small would be about the size of a sports stadium, like the Houston Astrodome. And if it were longish, it may be close to a 6 kilometer (4 mi) speed trap of a four-lane highway passing through a sleepy Midwest town.

Try cutting a glacier this small into drink-sized ice cubes. There would be enough for about 40 billion frozen daiquiris (assuming ten cubes per drink). Let's see, for my next cocktail party, I could invite 5 billion of my closest friends.

Many definitions of glaciers further stipulate that the ice develop out of recrystallized snow. Unfortunately, the criteria for birth from snow does not fit many popular kinds of glaciers, like rock glaciers and glacier remanié, which are described later in this chapter.

To distinguish glacier ice from other transient forms of ice that are typical of winter, there is often an age limit. Ice must be over one year old to be considered part of a glacier. Our definition of a glacier up to this point is a mass of ice that is larger than one-tenth of a square kilometer and older than one year. But there is more.

To Move or Not to Move

Whether or not an icy body is big or little or born out of snow, many believe that for it to be a real glacier it must show some type of movement. Ice naturally moves downhill under the force of gravity, so this should not be a problem. However, no one seems to agree about when the movement should take place—past or present—or how much movement is necessary for qualification as a glacier.

Evidence of past movement is common. Many of the remaining mountain glaciers, ice patches, and perennial snow fields in North America were once part of the massive glaciation that occurred between ten thousand and twenty thousand years ago. Geologists can confirm this by pointing to large amounts of geologic evidence showing past glacier movement. In this sense you could say that most ice remnants large enough to cover the Astrodome are glaciers, whether the ice is currently stagnant or not.

Those who contend that ice must show current movement in order to attain classification as a true glacier have a bit more of a problem. Movement is so relative that it is often difficult to identify. Many believe that glacier ice must move with sufficient force to pluck and scrape rocks away from the slope as it descends. Unless you can see the underbelly of a glacier, there is no sure way of telling if a glacier is sliding like this. However, there are some clues that help.

Little rocks held in ice act like sandpaper. When the rock-laden ice slides over bedrock, it scrapes away tiny fragments of rock (smaller even than a grain of sand) called glacier flour. The amount of flour created by this grinder mechanism depends upon the glacier's rate of movement and the type of underlying bedrock.

The flour is flushed out of the glacier through its meltwater drainage system, into rivers and lakes. It causes the water to look gray or milky as the little particles scatter incoming light. In a turbid river this milky color can persist for its full length downstream unless it is diluted by clear water from tributary streams. In lakes and oceans the water can develop a banded appearance from gray to turquoise as the flour gradually sinks. These bands can become very distinct, especially in ocean water where different temperatures and salinities maintain well-defined layering. The banded colors have been known to persist over 70 kilometers (40 mi) from a source glacier.

— *TRY THIS* —

Collect some water that is flowing directly from a nearby glacier, then drain the water through a coffee filter. The residue left in the filter is glacier flour.

It is possible to collect a large enough sample to see and compare. First weigh the coffee filter. Next pour 10 liters (3 gal) of glacier water through the filter (using a one-liter water bottle, this should require 10 dips and dumps).

Let the flour-filled filter rest in the sunshine to dry out. Now weigh the whole sample and subtract the original weight of the empty filter.

Depending upon the speed of the glacier, the volume of flowing water, and the underlying bedrock, you should typically find 100 milligrams to 10,000 milligrams (.004 to .4 oz) of flour in the 10 liters (3 gal) of water you collected. Sometimes the glacier flour content is as high as 500,000 milligrams (18 oz) per 10 liters (3 gal).

Rapidly moving glaciers can be more obvious. Usually there are cracks or crevasses present where the fast-moving ice pulls apart. Be aware, though, that seasonal snow can also develop some pretty deep crevasse-like cracks. Remember that a glacier must be composed of ice. So next time you see a crack, try to look down into it. If there is ice in there it will appear significantly bluer than the predominantly white snow and this could help you decide if you are looking into a glacier or not.

Do "Glacierettes" Have More Fun?

Although scientists cannot agree on a rigorous definition for glaciers, they have been able to group glaciers into distinct categories. Because of this we can picture a characteristic size, shape, or type of motion when a particular kind of glacier is mentioned.

There are two basic types of glaciers, those that flow outward in all directions with little regard for any underlying terrain and those that are confined by terrain to a particular path.

Snuggly Covers

The first category of glaciers includes those massive blankets that cover whole continents, appropriately called ice sheets. There must be over 50,000 square kilometers (10 million acres) of land covered with ice to qualify as an ice sheet. When cake-like portions of an ice sheet spread out over the ocean they form ice shelves.

About twenty thousand years ago the Cordilleran Ice Sheet covered nearly all the mountains in southern Alaska, western Canada, and the western United States (see Fig. 2.1). It was about 3 kilometers (2 mi) deep at its thickest point in northern Alberta. Now, there are only two ice sheets left on Earth, those covering Greenland and Antarctica.

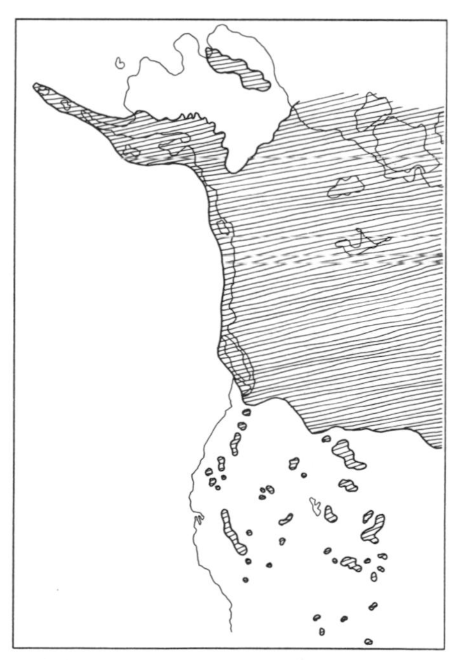

FIG. 2.1—The Cordilleran Ice Sheet once covered nearly all of the mountains in southern Alaska, western Canada, and the western United States.

Any dome-like body of ice that also flows out in all directions but is less than 50,000 square kilometers (10 million acres) is called an ice cap. Although ice caps are rare nowadays, there are a number in northeastern Canada, on Baffin Island and the Queen Elizabeth Islands, that protect the land from polar chill.

Alpine Blankets

The second category of glaciers includes a variety of shapes and sizes generally called mountain or alpine glaciers. Mountain glaciers are typically identified by the land form that controls their flow.

One form of mountain glacier that resembles an ice cap in that it flows outward in several directions, is called an icefield. The difference between an icefield and an ice cap is subtle. Essentially, the flow of an icefield is somewhat controlled by terrain and thus does not have the dome-like shape of a cap.

There are several icefields in the Wrangell, St. Elias, and Chugach mountains of Alaska and northern British Columbia. However, perhaps the most famous icefield is the Columbia in Banff/Jasper National Park on the border of British Columbia and Alberta. This and other icefields can be seen by driving along the aptly named Icefields Parkway.

Although less spectacular than a large icefield, the most common types of mountain glacier are cirque and valley glaciers. Cirque glaciers are found in basins or amphitheaters near ridge crests (see Fig. 2.2). Most have a characteristic circular shape, with the width as wide or wider than the length. Cirque glaciers average about 1 square kilometer (250 acres) and are usually less than 100 meters (300 ft) thick.

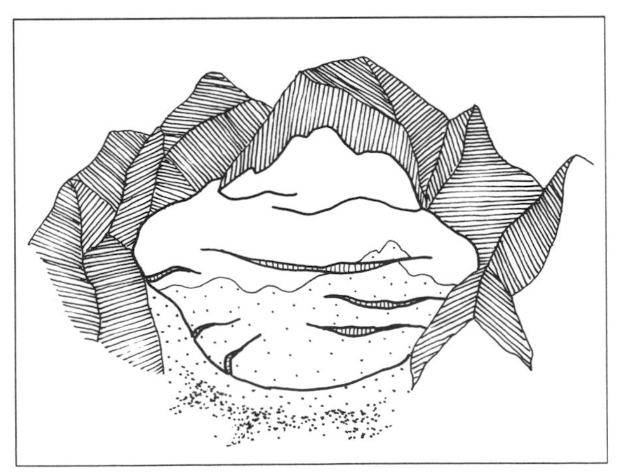

FIG. 2.2—Cirque glaciers are usually small, occupying a basin or amphitheater near ridge crests.

There are thousands of cirque glaciers in North America. Many are unnamed because they dot the landscape next to much larger ice bodies. Probably the best place to observe a cirque glacier is where it is regaled as the grandest ice around, as in the Sierra Nevada or the U.S. Rockies. One such cirque glacier is the Grinnell in Glacier National Park, Montana. Although small by most standards, the Grinnell Glacier is impressive, nestled into a steep-walled amphitheater next to the crest of the rugged Continental Divide.

Valley glaciers are the most suited to being called a "river of ice" (see Fig. 2.3.). Bounded by terrain the ice flows down valleys, curves around corners, and falls over cliffs. Although there are hundreds of short valley glaciers in North America, there are also many long ones. The longest valley glacier in North America at this writing is the Bering Glacier that flows off the Chugach mountains in southcentral Alaska. It cuts a 203 kilometer (126 mi) swath through the mountains and could be well over 500 meters (1,600 ft) thick in places.

FIG. 2.3—The long tongue of a valley glacier resembles a river of ice.

Many glaciers have tributaries that flow into them. These branched-valley glaciers can be very small with one or two unnamed tributaries, or they can be huge with several formidable glaciers flowing into one main branch.

If a valley or cirque glacier ends abruptly at or near the top of a cliff it is called a hanging glacier (see Fig. 2.4). Although quite beautiful and spectacular, hanging glaciers can be very dangerous if you walk underneath them. They frequently drop large ice chunks to the valley floor as the glacier pushes over its cliff.

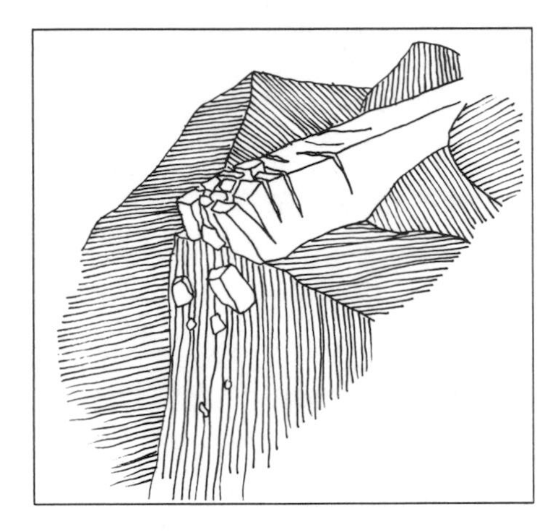

FIG. 2.4—When a glacier ends abruptly near the top of a cliff, it is called a hanging glacier.

Two hanging glaciers can be seen on the road to Portage Glacier in Alaska's Chugach Mountains. The Explorer Glacier and the Middle Glacier nudge toward cliffs that tower over the roadway. Both glaciers have recently receded back a little from their cliff edge so the danger from falling ice has been reduced. However, they remain spectacular, and there are photo opportunities along the way at each glacier vista.

Valley glaciers that flow into the ocean become tidewater glaciers (see Fig. 2.5). There are over fifty of these flowing into the Pacific Ocean from North America. All are breathtaking.

FIG. 2.5—Tidewater glaciers end in the ocean.

Certainly the most popular place to visit tidewater glaciers is Glacier Bay National Park west of Juneau in southeast Alaska. Another tremendous array of tidewater glaciers flows from the rugged Chugach Range into Prince William Sound in southcentral Alaska. Scenic flights, cruise ships, charter boats, and other guide services are making these areas increasingly accessible.

When mountain valley glaciers extend out over a plain and spread into a great lobe-shape pattern, they are called piedmont glaciers (see Fig. 2.6). A fantastic piedmont glacier can be found in southeast Alaska, the Malaspina. At over 5,000 square kilometers (1 million acres) the Malaspina is large enough to cover the state of Delaware.

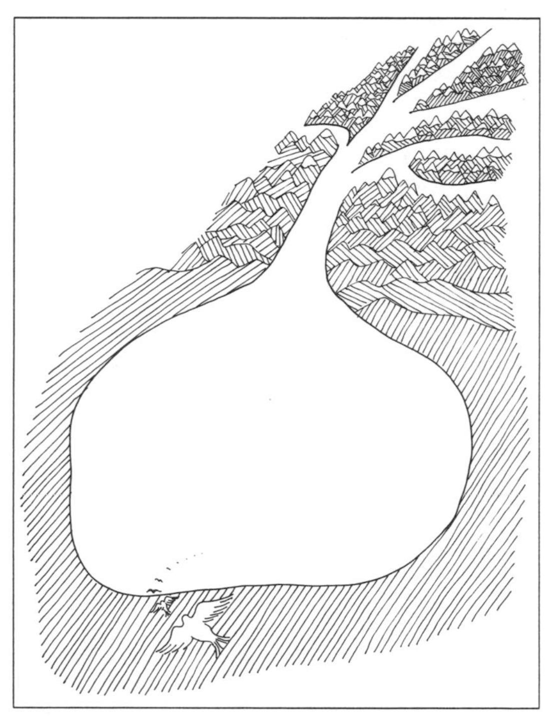

FIG. 2.6—The huge lobe-shape pattern of a piedmont glacier spreads out over a plain at the base of a mountain.

Although you can see the Malaspina by looking north across the harbor from Yakutat, the glacier is so large it is best appreciated from the air. You will recognize it quickly. It is the largest piece of ice that you can imagine. And that is not the half of it. The Malaspina is just a piedmont extension of the Seward Glacier, a mighty branched-valley glacier in its own right.

There is another kind of glacier that contributes to the towering landscape of the Canadian Rockies called a glacier remanié. These are reworked or reconstituted out of old glacier material. They develop at the base of a slope by ice blocks that accumulate and fuse together after avalanching from an overhead hanging glacier.

The Horseshoe Glacier in British Columbia is a remanié. The glacier obtains most of its nourishment by ice falling from above and relies little on seasonal snowfall. As a result, Horseshoe begins at an elevation well below the perennial snow line.

There are other weird little names for mountain glaciers. For example, ice aprons are thin sheets of ice that drape a mountainside, and niche or pocket glaciers occupy a small hollow or recess in a slope. Catchment glaciers are nourished by windblown snow from adjacent snow fields. Some people refer to these and the littlest cirques as glacierets, as they are too small to warrant the official title of glacier.

Rock, Rattle, and Roll

A unique form of glacier that is often left out of most basic texts on glaciers is a rock glacier. Although there is an ongoing debate about the definition of a rock glacier, it appears they can be categorized into three basic groups. The most common and perhaps the most true type of rock glacier is that formed in talus that is subject to permafrost. Another type is built by a collection of rock and snow or ice debris from avalanching. A third type is an old ice glacier simply covered by rock debris.

The talus-type rock glaciers can develop in any collection of talus that exists at a high enough elevation or latitude to be in an area of persistent permafrost. Talus is composed of coarse, angular rock fragments that lie at the base of a cliff or steep, rocky slope. The San Juan and Elk mountains of Colorado have all the necessary ingredients for rock glaciers—lots of talus and high elevation permafrost.

The rocks in talus glaciers are surrounded by ice that cements them together. These ice-cemented rock glaciers have the general appearance of a small cirque glacier with an odd bulbous shape. Their lower edge usually shows a series of arch-shaped ridges. When active, a talus rock glacier may be up to 50 meters (160 ft) thick.

Rock glaciers formed by avalanching material have much more snow and ice among the rock than talus-type rock glaciers. Therefore they look like and behave similarly to regular ice glaciers, except that they are black in color. The Wenkchemna Glacier in Banff/Jasper National Park illustrates this type of rock glacier. Ice, snow, and rock avalanche from several adjacent couloirs to collect in the main glacier.

Including debris-covered glaciers in the category of rock glaciers is considered blasphemy by many. Nonetheless, it is gaining popularity. This type of glacier is distinctive in that there is a core of ice underneath a blanket of rocks, much like the cream center of a Twinkie®.

Ice-cored rock glaciers have typically lost their original source of nourishment from newly fallen snow. Their only means of survival is the insulating cover of rock that prevents the remaining ice from melting away. This requires a rock blanket that is at least as thick as a good down comforter. The thicker the layer, the better it will insulate against daily, seasonal, and annual variations in temperature.

Usually the rock debris must be more than a meter (3 ft) thick to provide durable insulation. It accumulates on the glacier surface by avalanches, land slides, or melted-out moraine material. In a sense, these are dying glaciers, but they may still show signs of movement for years after burial.

The Shoestring Glacier on Mt. St. Helens is an impressive example of a debris-covered glacier or, if you prefer, an ice-cored rock glacier. The 1980 volcanic eruption blew its accumulation zone all over eastern Washington and dumped a load of rock and pumice on the remaining glacier surface. The rock covering insulated Shoestring from the subsequent heat and helped it to survive.

A much older ice-cored rock glacier can be found in Kananaskis Country north of Highland Pass alongside Alberta's Highway 40. It is believed to have formed in moraine that was built by glaciers over ten thousand years ago.

All Ice Is Not Cold

In addition to the morphological categories described above, glaciers are also distinguished by their characteristic temperatures. The amount of water generated by a glacier depends upon its temperature and the temperature of its surrounding environment; this means we can estimate such things as a glacier's contribution of water resources for drinking, irrigating, etc. The motion of a glacier depends upon the temperature of the ice as well, and the way a glacier moves can help determine a variety of characteristics, such as how it will respond to changes in climate.

A "warm" glacier is called a temperate glacier. The ice is at its melting temperature—remaining about 0° C (32° F) throughout, except for a thin layer near the surface that is cooled during winter. At the melting temperature ice and water can coexist without the ice melting or the water freezing. There is usually quite a bit of water running through a temperate glacier, especially during the summer. Since the majority of mountain glaciers in North America are temperate, this book is primarily about temperate glaciers.

Glaciers that are mostly well below freezing are polar glaciers. They may have a thin layer of melting temperatures near the surface during summer, or there may be a thin layer of melt at the bottom from frictional heating. Most polar glaciers exist in polar regions of the globe and many can be found on Canada's Arctic islands.

Transition glaciers that have both "warm" and "cold" ice are called subpolar glaciers. The Brooks Range in Alaska and parts of western Canada's northern interior have subpolar glaciers. The ice is generally cold, but there is extensive melt during the summer.

Classifying a glacier by its temperature is not entirely justified because there are many cases where different thermal patterns can be observed at different locations on the same glacier. However, temperature is an important part of the overall character of a glacier as we will discover in the following chapters.

The Great Worm Round-up

One way to determine if a patch of snow is perennial or overlying a true glacier is to search for the infamous ice worm. Aside from a few snow fleas, this is the only animal known to live in glaciers. They are related to the common earthworm, class oligochaete. The several species of ice worms range in color from yellowish brown to reddish brown, almost black. They are less than 1 millimeter (.04 in) in diameter, average about 3 millimeters (0.12 in) long, and live in snow, firn, or glacier ice (see Fig. 2.7).

These little worms are very sensitive to temperature. They exist best right near freezing, at 0° C (32° F). For this reason, they require a temperate glacier or thick patch of perennial snow where they can burrow deep enough to avoid the winter chill.

During summer, ice worms move near the glacier surface. However, it is too warm on sunny days for ice worms to exist right at the surface. They burrow down about 20 centimeters (8 in) to avoid the sun's heat and come up only after the sun sets and radiant heat diminishes. If you pick them up in your hand you can see how they squirm in the heat and try to escape downward.

Look for ice worms on cloudy days or after the sun has left a slope. They will often congregate around patches of red algae (another common feature of temperate glaciers and perennial snow patches). Many believe that red algae is a source of food for ice worms.

The red algae, which contains mostly water, resists freezing by incorporating minerals from dust in the snow, which help to depress its freezing temperature. This is the same process that prompts you to add rock salt to the ice-water bath surrounding an ice cream maker. The salt causes the ice-water mixture to be colder than the freezing temperature of pure ice. This allows heat to be drawn away from the ice cream.

I once overheard a conversation between a woman and her husband. She asked him why the snow was red, and he answered that the rangers had sprayed paint on the snow to mark the trail. Although completely wrong, this is a pretty fair description of what red algae looks like on the snow surface. Some people call the algae-covered surface "watermelon snow."

FIG. 2.7—Ice worm soup: Simmer one ton of firn 'til thick.

What's in a Name?

Based on these definitions, I would be hard pressed to call the St. Mary's Glacier in Colorado a true glacier. I found no evidence of ice. The water flowing from its melting body contained no glacier flour—the terminal lake was deep blue/green with no hint of a milky or gray color. There were no crevasses. Goodness, there were not even any worms.

Although too small to be counted into any true glacier inventory, there may have been some buried ice that I could not see. Therefore, at best, I might call the St. Mary's a glacieret. At worst, she is simply a patch of snow that clings to life year after year.

Despite the diminished dignity of a very popular playground, the St. Mary's should not lose all hope. A future ice age remains possible. By putting all the categories of glaciers into place you can imagine what an ice age would create. Glacierets would grow into full-sized glaciers. Cirques would develop into valley glaciers. The valley glaciers would join into icefields. The icefields would deepen and become ice caps. The caps would expand into ice sheets ... and the woolly-mitten industry would be booming.

CHAPTER
—3—

Crystals and Bubbles

I peered through my magnifying glass at a handful of snow. As I looked closer at the shapes and sizes, I spied a beautiful little crystal alone off to one side. It was perfectly shaped into a small, hexagonal plate with delicate spikes and spears flashing from each corner. The cold air kept the crystal edges sharp and clear as I studied its crisp lines and the teeniest, tiniest details I had ever witnessed. It was more perfect than anything that could last under the scrutiny of a camera.

For a long while I stared with reverence. Finally, under the warmth of my body heat the crystal reshaped itself, first losing its delicate spikes and spears into a simple plate, then rounding into a small ball. I fell back into the blanket of snow, absorbed in a rapture of beauty. To this day, millions of snow crystals later, I have yet to see anything quite like it.

My little crystal was less than .25 millimeter (0.01 in) in diameter, just about the size of a grain of salt. If I had left it in the snowpack it would have joined with its neighbors into a massive network of convoluted shapes, eventually growing and deforming into what we call glacier ice.

The process by which glacier ice is formed from snow crystals still intrigues many scientists. Obviously it takes more than one snowflake to make a glacier. Considering that there are over 50 million snow crystals in the size of a decent snowball, just imagine the zillions that are required to build a little glacier the size of a football stadium.

These newly fallen snow crystals are often delicate shapes of ice that nearly float in a bed of air. In fact, a layer of fresh snow is about 90 percent air! No wonder skiers have a sense of floating when they turn through a deep bed of powder snow.

The amount of mass in a given volume is called density. The density of new snow averages 100 kilograms per cubic meter (kg/m^3) (6 lb/ft^3), or 10 percent of the density of water. An average-sized suitcase would weigh about as much as a two-year-old child if it were full of fresh, low-density snow.

Soon after deposition, snow becomes denser rapidly as the crystals settle and pack close together under the force of gravity. Within a few hours to a day or two the density can double or triple in this way. But this is still a long, long way from glacier ice. For snow to densify further it must change shape, or metamorphose.

Névé Névé Land

The most common type of snow metamorphism reshapes the crystals into rounded grains that eventually bond together into simple chains. In the past, this type of recrystallization has been called destructive, equi-temperature, or equilibrium metamorphism. Because it depends upon the shape of a crystal, many refer to this type of metamorphism as radius-dependent recrystallization.

When a water molecule is precariously attached to the sharp end of a crystal, or convex portion, it has relatively few neighbors to hold onto. Without any significant attachments, it can leave easily. However, if a molecule is on a flat or hollow portion of a crystal, it has more neighbors to hold it in place and is less likely to leave.

On a delicately shaped snow crystal, molecules gradually migrate from sharp peaks toward the flats or hollows. This molecular movement causes the crystal's sharpness to fade, its delicate arms to round off, and the hollows to fill (see Fig. 3.1).

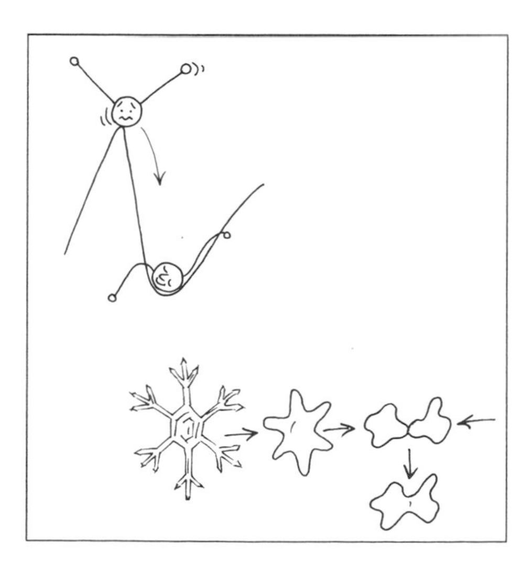

FIG. 3.1—Water molecules on an ice crystal leave sharp peaks and migrate toward flats and hollows. This causes the crystal to round and allows two touching crystals to bond together.

The place where two crystals touch also forms a hollow. Molecules migrate and fill in these hollows forming ice bonds between the grains. This is called sintering and chains of several grains commonly develop.

Once snow has close-packed, changed shape, and bonded together in this way it can reach a density of about 550 kg/m^3 (34 lb/ft^3) within a few weeks or months, depending upon the surrounding environment. For example, when temperatures are relatively warm, the molecules will be more energetic and the transformation can take place quickly, within a few weeks. If the temperature is cold, molecules are slow to move and the change can take months. As we will see later, the presence of liquid water can speed the whole process dramatically.

If this high-density snow survives at least one melt season on a glacier it is called firn or névé. Imagine your suitcase-sized sample of snow now to be comprised of firn instead of light snow. It would weigh about as much as a trim young woman, significantly heavier than our two-year-old.

A grain of firn may be a single crystal or it could be a group of crystals whose boundaries have grown big enough to be hidden in the rounded glob of ice. The grain itself is identified as a single unit that can be easily melted or broken out of the ice fabric. The diameter of a firn grain is usually between 1 and 4 millimeters (0.04 and 0.20 in) or about four to sixteen times larger than our original snow crystal.

Bubbles, Bubbles, Toil and Trouble

The next stage in developing glacier ice occurs slowly. As the density of firn increases, the spaces between grains grow smaller. This means that molecules wanting to evaporate from a crystal into the pore spaces have no room to move.

That is not to say that ice-locked molecules do not want to move. While the crystals have been metamorphosing, new layers of snow have been deposited. This creates an enormous amount of pressure on the underlying layers. Underneath this load the crystals try to reorient themselves to relieve the stress—just as the bottom man wriggles about in a wrestling match (see Fig. 3.2).

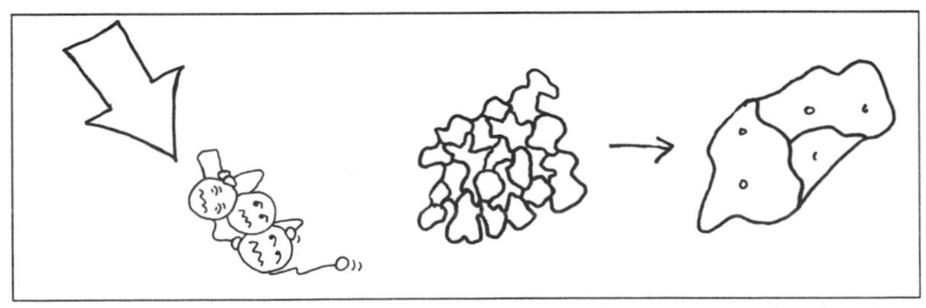

FIG. 3.2—Rounded snow grains that are pressed and squeezed together form large glacier ice crystals.

Instead of moving through the air then, many molecules bound to grains of firn slowly migrate over the surface or through the core of the crystal. After a while this slow migration squeezes off the air spaces entirely forming little bubbles in an otherwise solid ice structure. This is called glacier ice. It has a density of at least 830 kg/m^3 (52 lb/ft^3).

— Try This —

One way to determine if a sample of glacier is dense enough to be called ice is to check its air permeability, or how easily air flows through the sample. Glacier ice is impermeable to air. All of its trapped air is in isolated bubbles that cannot communicate with the surrounding atmosphere.

Find a fist-sized sample of glacier and hold it close to your lips. Now blow hard. If air comes back in your face, you probably have a sample of glacier ice in your hand. If you can blow through the sample, and perhaps tickle the ear of a close friend, you are probably holding some dense firn or snow. Be careful not to chill your lips so much that they turn blue.

At this point, a suitcase-sized sample of glacier ice would weigh as much as a heavyweight fighter. Unfortunately, a falling chunk of ice does not float like a butterfly, and it surely stings harder than a bee (see Fig. 3.3).

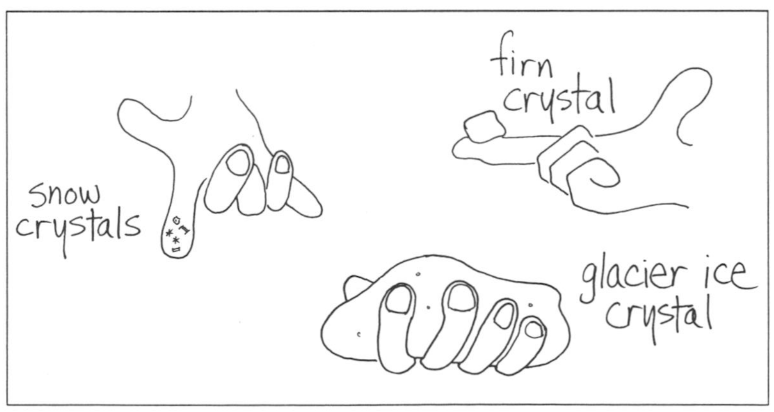

FIG. 3.3—Ice crystals grow dramatically as they transform from snowflakes and firn into glacier ice.

The time that it takes for firn to become ice depends upon temperature and the amount of loading. In the far northern regions of Alaska and Canada, where temperatures are cold and annual snowfall is low, it can take over a hundred years for glacier ice to develop. By the time it does so, the ice may be buried 50 meters (160 ft) or more below the surface. South of these areas, temperatures are warmer and annual snowfall is usually greater. Firn can change to ice within three to five years on the coastal glaciers of southeast Alaska, British Columbia, Washington, Oregon, and California. It may take ten to fifteen years for glacier ice to form in the somewhat cooler and drier regions of central Alaska, eastern British Columbia, Alberta, Montana, and Colorado.

Many times you can easily separate crystals with a little melting. As the crystals develop they push water soluble impurities like salts and gases out of the molecular structure, concentrating them at the crystal boundary. These impurities cause ice to melt at temperatures colder than its normal melting temperature, allowing the boundaries between crystals to melt faster than the clean ice within crystals.

When the boundaries between glacier ice grains are broken, either by melting or some external force, you might see crystals on a glacier's surface or lining the wall of a crevasse that are large enough to fit nicely into a tall glass of iced tea, about 3 to 5 centimeters (1 to 2 in) in diameter.

— *Try This* —

In the early 1900s Gerald Seligman studied grain structure in glaciers using rubbings, much like the tombstone rubbings done by artists today. Here is how he did it.

1. Find a smooth, weathered surface of ice where the crystal boundaries have melted slightly. These are most likely on walls of crevasses that have been exposed to warm air temperatures.
2. Wipe the ice gently with a pad of blotting paper to absorb some of the extra liquid water.
3. Use nonabsorbent paper, like a thin sheet with glossy surfaces for the actual rubbing. Keep the paper rolled, unrolling only long enough to rub a section so that it touches the ice only momentarily.
4. Use a soft pencil, crayon, or heelball (a solid dye used by bootmakers) and rub over the ice with an even pressure.

There is another, rather fascinating way to view the structure of grains and crystals from the inside out. Although much more subtle than the melt observed at crystal boundaries, shallow, parallel grooves can be observed on the face of a partly melted crystal. These are called Forel stripes. The grooves are parallel to individual planes of oriented molecules inside the crystal called basal planes. You can tell if a grain is made up of more than one crystal by observing the orientation of Forel stripes. Groups of Forel stripes aligned in more than one direction indicate there is more than one crystal.

Sometimes melting can be observed inside the crystal as thin, silvery discs, about 1 millimeter (0.04 in) in diameter. These internal melt features are called Tyndall figures after a famous British physicist who discovered them in the 1850s. The discs are parallel to the basal plane of an ice crystal, and their orientation can be observed in the same way as Forel stripes to determine the number and size of crystals in a grain of ice.

It usually takes direct sunlight to penetrate the crystal and create Tyndall figures. It helps to use a magnifying glass to focus the sunlight into the crystal. The sunlight will reflect off these melting planes and help you see them. Continued melting may allow the shapes to flower into a branch-like pattern of hexagonal symmetry. These are appropriately called Tyndall *flowers* and are like inverse snowflakes.

The isolated pockets of air trapped between glacier ice crystals that are squeezed and compressed into small bubbles can attain pressures that are over twenty times the amount of pressure inside an average car tire, or about fifty times that of standard atmospheric pressure at sea level. Just imagine what happens when they pop!

Alaskans have coined the phrases "ice sizzle" and "bergy seltzer" to describe the popping sound of bubbles that burst while icebergs are melting. I hear there is a tavern in the Canadian Rockies that serves specials on colored liqueurs whenever the local mountain guides supply glacier ice. Patrons peer into their glass, waiting for the ice bubbles to burst or the ice cube to crack along a line of bubbles. Each resulting display of spectacular colors brings a gleeful round of cheers (see Fig. 3.4).

FIG. 3.4—The air trapped inside the bubbles of glacier ice is often pressurized, sometimes more than twenty times the amount of pressure inside a standard automobile tire. The popping bubbles burst with great flare!

A distinguishing feature of glacier ice is its blue color. Clear glacier ice appears blue because it simultaneously absorbs light from the red end of the visible spectrum and scatters incoming light into shorter or blue wavelengths. However, if ice is bubbly, like firn, light will be quickly reflected out of the ice and cause it to appear white.

WATERING THE CRYSTAL GARDEN

The presence of liquid water accelerates the change from snow to firn. Also, refreezing large quantities of water can create ice lenses, making a rapid transition from snow directly to ice.

Because water transfers heat energy more rapidly than air, the molecules attached to wet ice grains can be energized to change location more frequently than if attached to "dry" ice. This causes the wet grains to round quickly. Also, because curvature still plays a vital role, small grains disappear as their molecular mass migrates toward large grains. The large grains get big fast and can reach sizes well over 1 millimeter (0.04 in) within a few hours or less.

Many times wet snow or firn grains will form clusters of several grains attaching to each other like the smooth bottom of a wet glass attaches to a counter-top. This type of recrystallization is called wet

— *TRY THIS* —

Sam Colbeck, a noted snow scientist, recommends a neat little experiment for those interested in wet metamorphism. He suggests that you collect some snow from your backyard (or use the frost from your freezer or shavings from ice cubes). Mix it with plenty of water that has been chilled to near freezing. Place the slushy snow into a small, sealed container.

Next, make an ice bath with a mixture of half water and half ice and submerge your sealed container of slushy snow into the bath. If possible, put the whole experiment in your refrigerator.

After 24 hours remove the sealed container and drain out all the water. Use some facial tissue to damp-dry the snow.

Reseal the container and place it back in the ice bath for a few hours.

When the experiment is over, pull everything out and look at the remaining ice with a magnifying glass of at least 5x magnification. You should see clusters of rounded ice particles, very similar to the structure of firn.

metamorphism. When the liquid water refreezes, it forms large, amorphous clusters of grains.

If the grains are saturated with water, the boundaries between grains become soaked. This slushy snow is very weak. However, when it refreezes a strong lens of solid ice develops.

There is another little-known type of crystal that can develop in watery depths of a temperate glacier. The crystal shapes range from large intricate branch-like flowers to short, stubby mushroom shapes.

These Thomson crystals grow in cavities, like those created by crevasses and moulins, that are partially filled with stagnant water. Although ice and water can coexist for long periods of time under the right temperature and pressure conditions, any subtle change can upset this balance.

In the case of Thomson crystals, the large amount of pressure exerted on ice that is far below a glacier's surface causes it to melt at temperatures that are slightly colder than its normal freezing point at typical atmospheric pressures. Heat flows outward from the slightly warmer water in the crevasse to the slightly colder surrounding ice, causing some of the crevasse water to freeze. The molecules refreeze into intricate patterns of Thomson crystals.

It can take months or years to grow these crystals, some of which develop over 20 centimeters (8 in) in diameter. Therefore they are rare and only found at great depths in relatively stagnant portions of glaciers.

— Try This —

When you are on a stagnant portion of glacier ice, look for an old, deep crevasse or moulin with some water in it. Drop some rocks down the side of the crevasse to dislodge any existing Thomson crystals. They will float to the surface like lily pads on a pond. They are quite pretty.

Through the Looking Glass

In addition to the ice lenses and crusts that are created through a melt-freeze cycle associated with wet metamorphism, there is another type of ice layer that can turn a glacier into a huge reflecting mirror. This ice sheet forms on a glacier's surface and strongly reflects sunlight. It makes the glacier look like it is on fire and for this reason it has been called "glacier fire." However, it is more commonly referred to as firnspiegel, a German word meaning firn mirror. (This can be somewhat misleading since firnspiegel can form over snow as well as firn.)

Firnspiegel forms under just the right combination of cold air and intense sunshine; this is usually at high altitudes and especially during spring. Solar radiation penetrates a snow or firn surface, melting the ice there. The meltwater, instead of filtering down through the grains below, will encounter the cold, dry air and refreeze on the surface. This creates a very thin sheet of clear ice.

The ice sheet of firnspiegel forms a glass-like covering over the melting material. Much like the effect of glass in a greenhouse, heat is trapped underneath. This helps to further melt the underlying snow or firn and a gap is formed under the overlying ice sheet.

Other ice sheets that appear similar to firnspiegel may form by freezing rain. These can be distinguished from firnspiegel because they are usually thicker and probably will not have the characteristic gap between the underlying snow or firn and the icy surface.

HOAR D'HOAR HOAR

When there is a variation of temperature across a snow layer another type of recrystallization takes place. It has been called constructive, temperature-gradient, and kinetic-growth metamorphism.

Most temperature variations occur when there is a strong contrast between cold air temperatures and relatively warm temperatures under the snow. They are strongest within the early winter snow when the glacier surface is still warm from the summer heat and the overlying new snow is shallow.

The differences in temperature cause molecules to evaporate from warm crystals and reconnect onto cold crystals (see Fig. 3.5). This movement can become quite rapid and when evaporated molecules slam against another crystal, they will attach themselves to a step in the crystal lattice. As the step builds out farther, its protrusion will entice other incoming molecules to join.

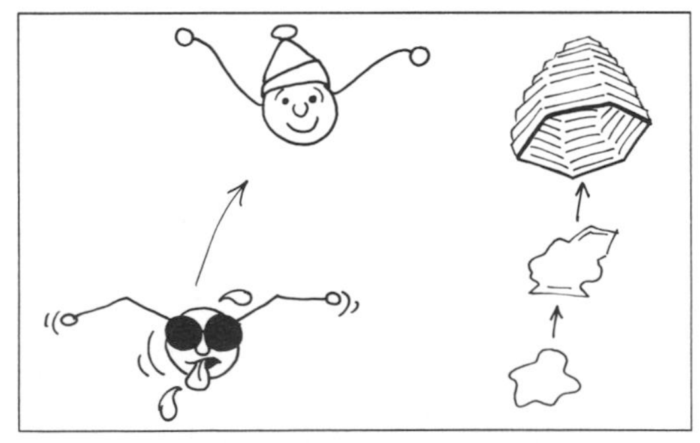

FIG. 3.5—When strong differences in temperature occur across a snow layer, water molecules leave warm areas and migrate toward cold areas. This causes ice grains to recrystallize into angular shapes like cups and scrolls.

You can actually see these steps as they develop on a crystal. The first sign of this type of rapid recrystallization is angular edges on the grain faces. If you had a teeny, tiny protractor and could measure the angle where two facets meet, you would probably find that the angle was about 120°. This is the same angle as occurs between two hydrogen atoms in a water molecule that is frozen into a crystal lattice, and not by coincidence.

With further growth of these facets, the steps become more apparent. Although a step may actually be only 1 molecule thick, the ones that are macroscopically visible are several thousand molecules thick. Faceted grains and some step development can occur in a wide variety of density and temperature conditions.

If the recrystallization process continues within a relatively low-density snowpack and with strong temperature gradients, a complete process of recrystallization will cause a stepped pattern of cups and scrolls. Many times the grains grow to several centimeters in diameter. At this stage the newly created grains are called depth hoar.

Once the temperature gradient is removed, faceted and stepped crystals can metamorphose slowly back to a rounded shape. However, the stepped pattern of depth hoar is one with very little variation in curvature. Without any radius-dependent forces (or water!) to cause metamorphism, the cup and scroll shapes can remain for long periods of time.

In addition, because the temperature gradient is usually vertical through the snowpack, the resulting grains most often recrystallize into vertically preferred shapes. This vertical growth pattern of depth hoar causes it to be resistant to the vertical forces of gravity, just like a tower built out of playing cards.

Because of its durability under temperature and pressure changes, persistent depth hoar will often create a distinctive layer within the glacier. In fact, many polar glaciers develop depth hoar each autumn that remain visible for years after being submerged.

Layer Cakes

Layers of the various crystals described above, alternating with deposits of dust, pollens, and volcanic ash, create a birthday cake-like pattern on a glacier. These layers not only are pretty but also can provide interesting clues to past climates, especially when they are associated with the annual cycle of seasons.

During the winter season, the snowpack is in a continual process of change. From the moment snow is deposited it undertakes the task of settling, condensing, and metamorphosing. By the time the last snowflake falls, most of the individual layers of snow that were deposited by unique storms have melded together into one cohesive group of grains called an annual layer.

During the summer, when much of the snow has melted away from surrounding rocks, wind blows the exposed dust and dirt onto the glacier surface. This dirty surface makes a distinctive layer well after being covered by a new winter's snow and can mark the boundary between annual layers (see Fig. 3.6). Although a dirt layer can remain distinctive for many years, in temperate glaciers the dirt particles may wash through the glacier firn as meltwater percolates between the pore spaces, making the dirt horizons less visible.

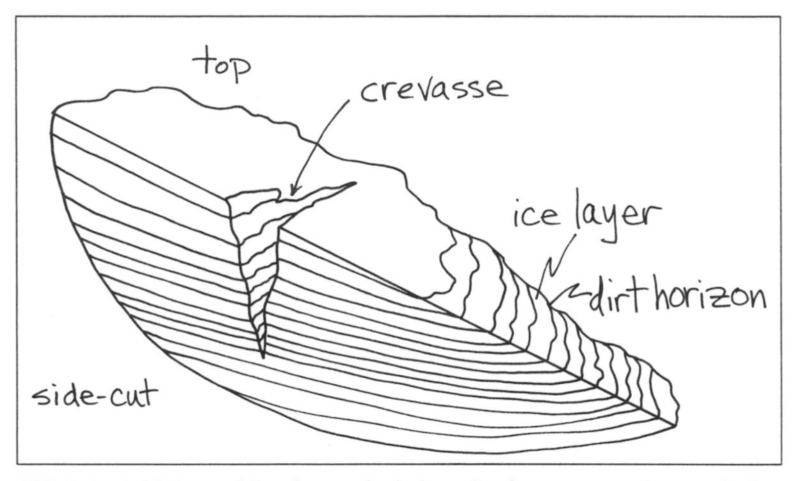

FIG. 3.6—A thin layer of dirt often marks the boundary between annual accumulations of snow and firn on a glacier. On large glaciers the best way to observe annual layers is to look at the wall of an open crevasse near the head of the glacier. On small glaciers, annual layers may be visible near the snout during the ablation season.

Volcanic eruptions also provide visible horizons in glaciers by depositing layers of ash on the exposed ice surface. Because ash is usually composed of particles larger in size than average dust, they can persist for longer periods, even in temperate glaciers.

Scientists also detect trace elements from natural causes, pollution, and nuclear bomb tests that are deposited on glaciers. These measurements can be used to monitor pollution levels and trends when compared with trace element concentrations in glacier layers that were formed before the industrial era. The recent nuclear reactor accident at Chernobyl has deposited radioactive material on many glaciers throughout the world. This will create a distinctive layer that can be monitored for many years to come.

A very unique type of layering can be seen on several small glaciers near the Beartooth Plateau of Montana. It seems that the migration flights of local grasshoppers coincide with heavy thunderstorm activity and snowfalls that are common in August and September. The severe weather causes grasshoppers to crash into the glaciers and subsequently freeze. Whole grasshoppers and grasshopper parts can be seen melting out of the ice each summer.

Many times these annual layers retain distinctive features even after they have developed into glacier ice. For example, depending upon the speed of transformation, one layer may have trapped more air bubbles than adjacent layers. This causes the bubbly ice layer to look decidedly different from its clear ice neighbors.

— *Try This* —

Crevasses open a hole to the inside of a glacier and they are great places to observe stratigraphy. Next time you are near a crevasse at the head of a glacier, look closely at one of the walls. See if you can identify distinctive annual layers of firn. Determine how long it took for firn to become ice by counting the annual layers from the surface to the firn and ice horizon.

A thin layer of firn between dirt horizons indicates a year of sparse snowfall and/or a heavy melt season. A thick layer indicates heavy annual snow and/or a cool summer. Can you determine something about the climate over the past few years? Are there layers of depth hoar or ice lenses that you can follow from one crevasse to another?

Stress and Folia

Foliation is another type of layering pattern found in glaciers. It occurs when glacier ice experiences a variety of stresses or changes in weathering patterns that cause crystals and bubbles to preferentially orient themselves. Weather-caused foliation is most common near icefalls, and we will talk about that in the next section.

Stress-induced foliation is most common near the valley walls and the glacier bed surface where the ice shears past stationary bedrock. Ice crystals usually change shape and reorient themselves to accommodate any applied loads. This causes stress-induced foliation to develop distinctive crystal sizes. The largest crystals formed this way are usually found where the ice had once experienced great stress, but current stresses are small—such as near the terminus of a glacier.

Coarse-grained ice is greater than 5 millimeters (0.2 in) and usually about 5 centimeters (2 in) in diameter. However, warmer temperatures and old age allow crystals to grow larger. Some glacier ice crystals in the huge Malaspina Glacier of Alaska are over 20 centimeters (8 in) in diameter.

Fine-grained ice is less than 5 millimeters (0.2 in). In very cold ice (below about -10° C [14° F]) the crystals never grow beyond 2 millimeters (0.08 in). It looks like snow when it is disaggregated.

The number of bubbles apparently influences the ultimate size of stress-transformed crystals and helps to distinguish glacier ice layers that contribute to foliation. For example, coarse-bubbly ice has grain diameters 1 to 6 centimeters (0.4 to 2.4 in). It looks whiter than most other ice because it is filled with small bubbles. It is the most common type of glacier ice and is often seen in ablation areas.

Coarse-clear ice has grain diameters slightly larger than coarse-bubbly ice at 3 to 12 centimeters (1 to 5 in). It is free of bubbles and can be found most often near the margins and terminus of a glacier. This is the bluest ice of all.

Any bubbles within glacier ice will get squeezed and reoriented under the same stresses that grow and reorient the surrounding ice crystals. It is not uncommon to find a once circular bubble that has been squeezed into a long rod or flat plane.

— *Try This* —

Even if you do not have immediate access to the highly pressurized bubbles found in glacier ice, you can see different bubble concentrations and shapes by looking at a cube of ice grown in your freezer.

Ice cubes in your freezer will freeze first on the outside, trapping air bubbles toward the center. This causes a bubble-free exterior and a bubbly interior with elongated bubbles in between.

No Jive, It's an Ogive

Ogives are alternating stripes of light and dark color on the glacier surface. The bands are caused by weathering of the foliation pattern.

There are many types of ogives. The most commonly studied are ogives that form at the base of icefalls. These types of ogives are most apparent below steep, narrow icefalls.

During summer the ice surface melts and collects windblown dust (see Fig. 3.7a). The opened crevasses will also fill with meltwater and dust. At the bottom of the icefall, the summer layers look dark with fine- and coarse-clear ice and an accumulation of dust.

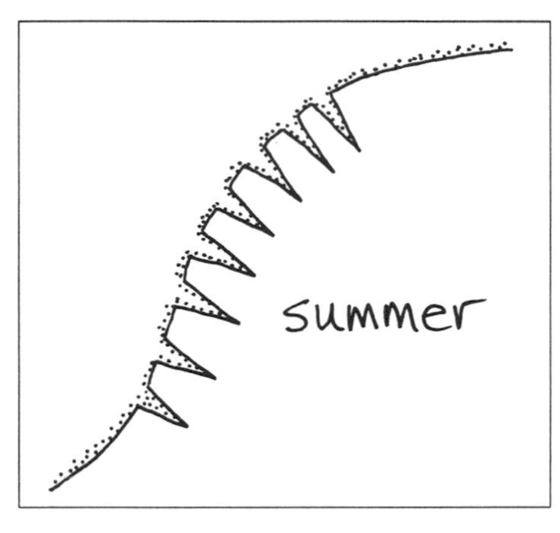

FIG. 3.7a—During summer, the ice surface melts and collects windblown dust and particles. This creates a dark layer of dirty, clear ice at the base of an icefall.

Winter snowfall protects the ice from weathering (see Fig. 3.7b). Open crevasses will fill with snow. These factors combine to cause layers reaching the bottom of the icefall during winter to remain white with coarse- and fine-bubbly ice.

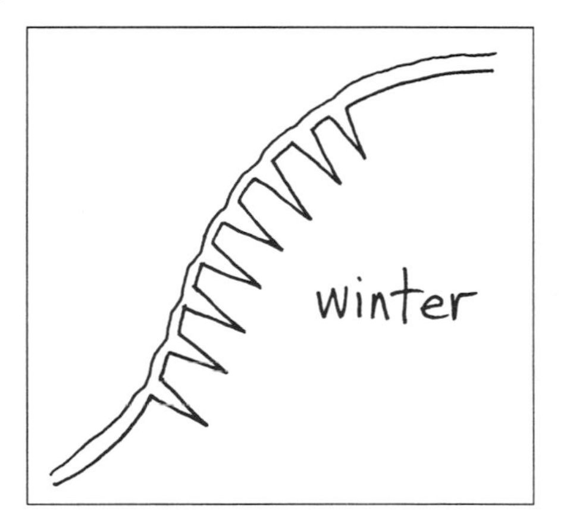

FIG. 3.7b—During winter, ice is covered with snow and protected from weathering. This creates a light layer of clean, bubbly ice.

The combined width of one light and one dark band corresponds to the amount a glacier travels within one year. In fact, the historical speed of glaciers has been calculated by measuring the width of a light/ dark ogive couple and comparing it with measured speeds during an ogive formation. The dark and light band ogives are called Forbes bands or band ogives.

Depending upon the length and steepness of the icefall and the duration of the ablation season, waves can form in the ogives (see Fig. 3.7c). The light-colored winter bands usually correspond to the wave crests and dark-colored summer bands correspond to wave troughs. The crest to trough amplitude on a glacier surface can be up to 10 meters (30 ft). Surf's up!

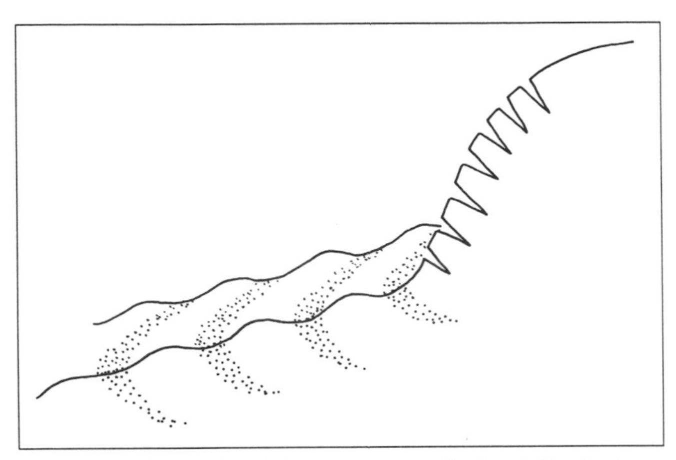

FIG. 3.7c—Wave or banded ogives are usually found directly down-glacier from steep, narrow icefalls.

Often only ten to twenty wave ogives are visible below an icefall. Other times there are many that persist a long distance down-glacier. In a few cases, the waves dampen so quickly that little if any surface undulation is observed.

Sometimes a wave-like band is created below an icefall by a collection of serac blocks. Rapid seasonal movement may cause more seracs to fall during summer and give these collections an annual band-like appearance. However, because the blocks are deposited on the glacier surface, they should disappear quickly in the ablation zone and be distinguished from ogives.

The easiest banded ogives to observe in North America seem to be on Mt. Rainier in Washington. The Cowlitz Glacier has also been noted for its ogives. However, these ogives are formed from an icefall off the Ingraham Glacier, which eventually flows into the Cowlitz. No matter whose ogives they are, they can be seen from the Cowlitz Rocks, just a short hike from the Paradise Visitor Center.

— *Try This* —

Next time you see an icefall, look for ogives from a high viewpoint. Are there bands of light and dark ice below the icefall? Check the relative spacing between couples of light and dark bands. A wide couple of light and dark would indicate faster annual movement than a thin band of light and dark. Are there years when the glacier moved more slowly than others?

Another type of ogive may develop below avalanche chutes. In this case, the pattern of light and dark reflects the frequency of avalanching rather than movement of the glacier. A snow avalanche causes frictional heating of the snow particles as they rapidly descend. This allows the resulting firn and ice to be coarser and less bubbly than a layer of precipitated snow. This, along with some entrained dirt and rock in the avalanche debris, may cause the avalanche-deposited layer to be darker than the surrounding ice.

Many times an avalanche is comprised of only rock and no snow. In this case the ogives created of rock and rock-free bands are often called "false ogives" because there is no foliate pattern of different crystal fabrics.

Ogive-like features also are created by annual dirt layers that become visible on some glaciers. These sometimes are called sedimentary ogives and are described in Chapter Four.

Cracks Are Folia Too

Just like in icefalls, the cracks of crevasses on other parts of a glacier can fill with snow or water and create a distinctive foliation. When filled with snow, the resulting ice band will be bubblier than surrounding ice. When filled with water, the refrozen liquid will be less bubbly than the surrounding ice.

Marching Bands

Because layer patterns on a glacier develop and deform as the glacier ice flows, finding the best place to observe them can be tricky. For example, annual layers on large glaciers are usually distorted before they reach the terminus. Therefore it may be best to peer inside a crevasse near the head of a glacier to observe its annual layer patterns. On the other hand, small glaciers that experience only simple stresses may still have their annual layer patterns visible at the glacier terminus.

Stress-caused foliation patterns can be found near the margins and termini of large glaciers. The most dramatic patterns are found on glaciers that have a tortuous path between head and toe. You should see alternating bubbly and bubble-free and coarse- and fine-grained foliation patterns that are caused by complex stresses. Ice caves are fantastic places to see layers of bubbly and bubble-free ice, whereas the weathered surfaces of an ablation zone will reveal distinctive coarse- and fine-grained layers.

CHAPTER
—4—

A Snowflake a Day Keeps the Doctor Away

We were lying naked in Moose Piss Slough. It was shallow and murky. We lay flat, grabbing little roots poking through the mud bottom to keep ourselves under the soothing water. This was it. This was perfect. There was nothing else I could conceive of at the moment that could carry me to such heights of ecstasy.

We had come off the ice early that morning then battled our way through scrub alder as the sun rose higher and hotter. My sweat had crystallized, graying my skin and whitening my clothes. The elephant I seemed to be carrying in my pack had started shifting and poking me.

Moose Piss wasn't really a slough. But it wasn't a pond either. It was more like a big mud puddle. I suppose the glacier had helped to carve this little niche when it was down-valley this far before. During those times, I heard, the way off the ice had been much easier. Just my luck to catch it in its shrunken state.

In the serenity of Moose Piss Slough we began chatting about the health of the glacier we had just visited and the many others surrounding us. We talked about how they grow and shrink, some dying, others exhibiting a joie de vivre.

Although glaciers often seem to be unchanging rock-like masses of ice, if you watch one for a few years you will be amazed at how much it fluctuates in size and shape. Sometimes a glacier responds directly to shifting climate by growing or shrinking. Other times glaciers are simply adjusting to fit the surrounding terrain or some internal change in their structure.

Scientists monitor the size of glaciers and ice sheets to help determine the effects of global warming. The amount of ice held in glaciers also influences the rise and fall of our sea levels.

Many cities and towns across North America monitor the size of local glaciers because they contribute to the area's water supply. Tidewater glaciers in Alaska are studied because they can play havoc with shipping lanes and fishing grounds when they adjust their length. Even backcountry travellers pay keen attention to the changing shape of glaciers that often provide the only access to remote areas of the mountains.

Win Some, Lose Some

A glacier remains healthy by accumulating as much or more material than it loses. There are a few "gas guzzlers" in North America that accumulate huge volumes of snow each winter and lose equally

— *Try This* —

While in the mountains next spring, measure the amount of snow that has accumulated over the winter. It is best to wait long enough after the last snowfall for it to settle and consolidate, but before it begins to melt away.

One way to select the right time for your observations is to wait for at least three consecutive days and nights when temperatures remain above freezing at the glacier's location. This will usually be sometime in April, May, or June, depending upon the latitude, altitude, and prevailing weather.

Climb up to the elevation of your favorite glacier and measure the amount of snow that has accumulated during the winter by poking a long stick or rod into the snow until you hit the ground.

Unless you know how to safely travel on a glacier's névé, I suggest you confine your measurements to slopes that are adjacent to the glacier. However, be aware that snow resting on the uneven, dark surfaces of vegetation and rock will melt away faster than snow lying on the smooth, cold surface of glacier ice. Therefore, this measure of accumulation next to a glacier is only an estimate of the snow actually available for glacier growth.

If you do this at the same spot every year, you will begin to distinguish from fat and lean accumulation seasons. This will help you estimate how well the glacier has prepared for the upcoming melt season.

huge amounts during plentiful hot, summer days. There are also some economy glaciers that gather little or no nourishment but survive quite nicely because they lose so little.

Snowfall is the most important kind of accumulation on a glacier, but freezing rain, avalanches, wind-drifted snow, and material from tributary glaciers also can contribute. This causes the pattern of accumulating material to be much more intricate than the even coat of frosting on a cake. Indeed, because most mountain glaciers are surrounded by complex topography with complex weather patterns, the influx and distribution of material is widely variable over the surface of a glacier. In addition, the pattern of accumulation can change significantly from month to month, year to year (see Fig. 4.1). For this reason most estimates of glacier accumulation use indicators that are more or less related to the actual amount of accumulating material.

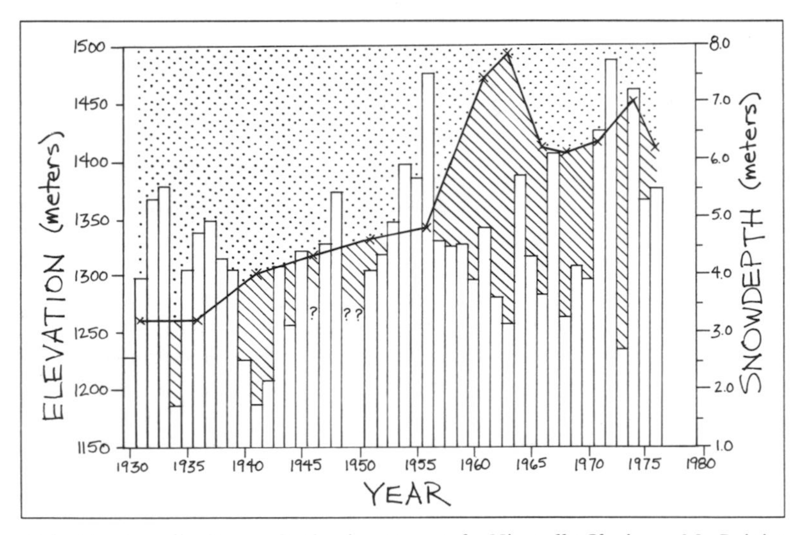

FIG. 4.1—Paradise Ranger Station is very near the Nisqually Glacier on Mt. Rainier. The snow that is left over on the ground in April can be compared with the elevation of Nisqually's terminus to show that Nisqually's health is somewhat dependent upon the accumulation of snow each winter. For example, look at the 1960s. During a period of below average snow depth, the glacier terminus retreated quite dramatically. (It is difficult to see an exact correlation here because snow depth was measured almost annually, and the snout elevation was measured about every decade. Also the amount of cloud cover during summer would help to estimate how much ablation the Nisqually had to suffer.)

A glacier loses material in a variety of ways. The wastage of a glacier, or its combined loss of ice, is called ablation. Much ablation occurs at the glacier surface and is caused by melting, with some evaporation and wind-scouring ablation. Like accumulation, ablation patterns can be variable. However, the amount of surface ablation seems more dependent upon the character of the existing snow or ice surface itself than complex features of surrounding terrain.

On a hot summer day many glaciers have been known to melt as much as 10 centimeters (4 in) of surface snow or firn. Over the surface of a small glacier, say 0.1 square kilometer (25 acres), with a cover of mostly firn, this would equate to about 6 million kilograms (13 million lb) of material. When melted, this amount of lost water would be enough to fill about three Olympic-sized swimming pools.

— Try This —

Next time you spend a summer day near a glacier, measure the amount of ablation due to melt. During the morning make an ablation-measuring stake by jamming a stick vertically into the glacier surface.

It is best to have a white stake so it will absorb as little heat as possible. A sun-bleached stick from a nearby forest should work fine.

Be sure that the end of the ablation stake is buried at least 15 centimeters (6 in) into the glacier surface so it will not melt completely out. After you have the stake securely stuck in the glacier, measure the distance between the top of the stake and the glacier surface. If the surface is rough, place your ice axe horizontally across the base of the ablation stake to help determine the average surface level.

When you return that night, measure the distance between the top of the stake to the glacier surface again. The difference between your current measurement and the morning measurement will tell you how much the glacier lost during the day.

If you really like this sort of thing, you can place stakes in various parts of the glacier to measure the difference in ablation that occurs at exposed ice surfaces and surfaces covered with snow or firn.

Because ice is more dense than snow or firn, the same amount of heating would cause a melting ice surface to diminish more slowly. That means that if your ablation stake measured a 10 centimeter (4 in) drop in the level of a firn-covered surface, then the same amount of melting may cause an ice surface to drop only about 6 centimeters (2 in).

However, solid ice surfaces can absorb more heat than surfaces of snow or firn because less solar radiation is scattered back into the atmosphere. Therefore, on any given day, the amount of melt on an ice surface could be three or four times greater than the amount of melt on a snow surface.

My Cup Runneth Away

Those of you who have spent time summer skiing on a glacier will have grumbled, at one time or another, at the often rough surface (see Fig. 4.2). The dimply surface is caused by subtle variations in ablation that are commonly referred to as sun cups. Because these cups are not always caused by the sun, they are more appropriately called ablation hollows. We can just call them cups for short, since that is what they look like in their beginning stages on most North American glaciers, a counter covered with thick-walled industrial coffee cups waiting to be washed. Typical ridge to hollow relief is 2 to 50 centimeters (1 to 20 in).

FIG. 4.2—Ablation hollows often form a rough dimply surface that can be difficult to negotiate when refrozen.

It is a little uncertain how cups begin forming. The honeycomb-like pattern of melt, typical of cups, may develop by variable patterns in debris, travelling sun rays, or even variable wind patterns that pump different amounts of heat to the surface in different places. Whatever the initial cause, a subtle pattern of ridges and troughs soon develops.

Further growth of cups is a very controversial issue. Some believe that cup development is enhanced by warm winds. Others believe that intense sunshine is required. At least one theory suggests that warm winds enhance cup development in dirty snow while sunshine enhances cup development in clean snow.

Consider a dirty, undulating snow surface that is being heated by warm winds. As the surface melts, debris particles that maintain their relative position on the melting surface appear to migrate toward the ridge axis of a sloping cup wall. When the snow surface melts further, there will be a greater concentration of particles along the ridges than in the hollows (see Fig. 4.3).

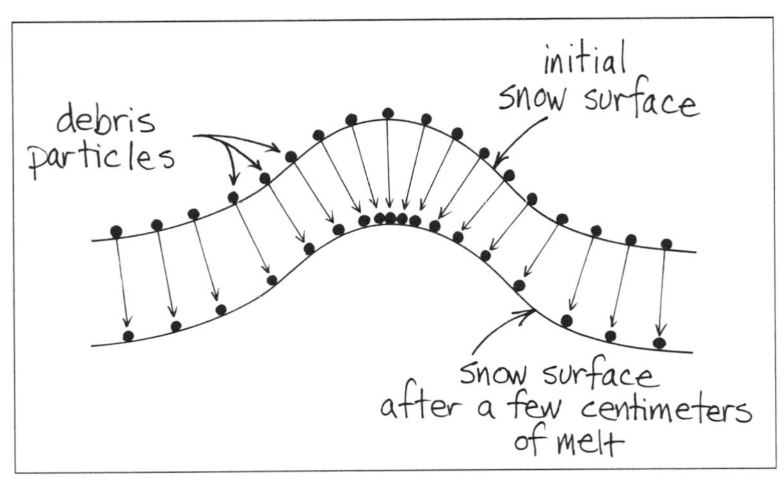

FIG. 4.3—Dirt particles resting on an undulating surface congregate on the ridges as the surface melts down.

Any amount of debris covering, whether dark or light, will help protect snow from wind and its associated "heat pump." This causes the turbulent exchange of heat to decrease over the debris-covered ridges and inhibit melting there. Meanwhile, melting in the clean hollows continues. The difference in melting rates enhances the cup's ridge-to-hollow relief.

Sunshine on this dirty snow surface will degrade the cups. Because the dirty ridges will absorb the sun's rays faster than the clean hollows, further sunshine will help to smooth the surface.

On clean snow sunshine will enhance cup development. Cup depths usually increase with elevation where the sun's rays are intense, but the air is relatively dry. A little bit of dry wind whisks away water molecules that are freed by solar heating from the ice lattice in the exposed ridges, causing them to evaporate. Meanwhile, water molecules that are heated to freedom in the protected hollows stay near the surface, forming a liquid layer that helps the ice surface to further melt. Since the rate of evaporation is significantly slower than the rate of melting, the melting hollows deepen dramatically in relation to the slowly evaporating ridges.

As the cups deepen they appear to migrate a few centimeters each day. I am not kidding: have cup, will travel. In the northern hemisphere this migration proceeds northward as the sun-exposed south-facing walls melt faster than the north-facing walls.

An extreme relief can occur between melting hollows and evaporating ridges that no longer resembles cups. The ridges become spike-like structures that look like the hooded figures of people requesting forgiveness for their sins (see Fig. 4.4). For this reason they are called penitents.

FIG. 4.4—Very deep ablation hollows resemble the hooded figures of penitent souls. These are usually found at high elevations on glaciers in the lower latitudes, like the volcanic peaks of southern Washington, Oregon, and California.

Penitents up to shoulder depth have been observed at high elevations on the glaciers of Mt. Shasta in northern California. However, they are more common in the high-elevation tropics where they are often over 3 meters (10 ft) high with hollows melted through to exposed ground. In either case, the spikes align toward solar noon, often growing together into east-west oriented lines to look more like marching soldiers than repentant souls.

Cone Heads and Funny Tables

As stated earlier, a light amount of dust helps to absorb the sun's heat and enhance melting. But there is such a thing as having too much of a good thing. When the debris becomes too thick, it will begin to insulate the glacier from the sun's rays.

It seems that the critical thickness for which a layer of dust particles becomes an insulator to direct sunshine is about 3 millimeters (0.12 in). Anything thicker than this will cause the underlying snow, firn, or ice to melt more slowly than the surrounding glacier.

Cone heads (actually called debris cones) form in this way (see Fig. 4.5). A pile of debris will insulate a small portion of the glacier as the surrounding, clean glacier material melts away. Eventually the slope between the debris-covered and non–debris-covered surface steepens and a cone shape forms.

As a debris cone grows the debris can slide off into surrounding depressions. The exposed ice melts away and the surrounding depressions become cones. This reverses the topography.

— TRY THIS —

A fellow named Alfred Jahn has been experimenting with the idea of cup formation using blades of grass or matchsticks to create what he calls "sun spirals." He has noticed that when sticks are placed on a snow or firn surface they will turn with or opposite to the sun depending upon cloud cover and time of day. In fact, if cups develop, the sticks will align along the ridges and form concentric rings. He uses these observations to show that cup formation and resulting debris patterns can be caused by the direct influence of the sun's rays. You should try it and see what you think.

Wait for a sunny day. To precisely observe the turning effect, use eighteen sticks and place them on a smooth snow or firn surface in a six-pointed star-like pattern, three sticks in each branch, with a small space between each, radiating from a central point.

The end of a stick that is warmed first, probably the one pointing closest to the direction of the sun, will sink into the snow. As time passes the sun will change its angle of incidence and travel from east to west. At the sun's later position, if the exposed end remains oriented away from the sun, then the stick will turn toward its already melted hole and rotate in the opposite direction as the sun, that is west to east. However, if the sun's new position catches the exposed end pointing toward it, then that end will be warmed and the stick will begin turning with the direction of the sun.

On a clear, sunny day it should take about three hours to begin noticing a turn. After twenty-four hours the pattern of sticks or grass will have changed into a "sun spiral." Note that partly cloudy skies will affect the direction of turn by causing the elements to sink at different times.

Large debris cones are often found near the edges of a retreating glacier made by dust, dirt, and rocks that had filled an old crevasse. When the surrounding glacier melts away, ice underneath the dirty crevasse is insulated and preserved. After the glacier has nearly vanished from the area, the debris-covered ice remains. The resulting debris cones look like small Egyptian pyramids. If you scrape off the debris you can see old glacier ice underneath.

Ice can become insulated underneath moraine material as well, forming hummocky shapes on, or near, the edge of a glacier. These ice-

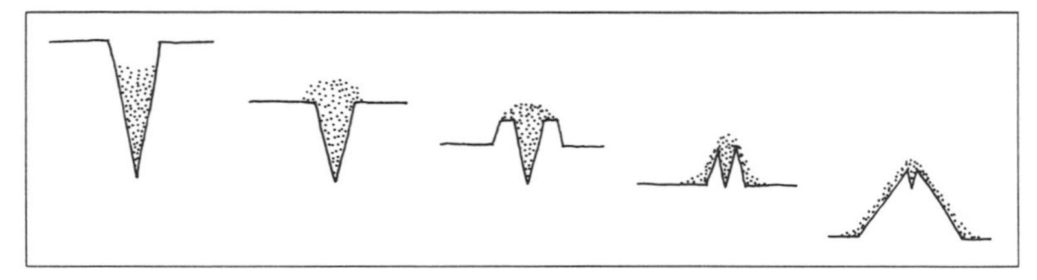

FIG. 4.5—When a glacier surface is covered by a thick coating of dirt or debris, it is sheltered from sunshine and melts less rapidly than the exposed ice that surrounds it. This can cause cones or pyramid shapes to form as the surrounding clean ice melts away

cored moraines and large dirt cones are potentially treacherous. They look like simple mounds of talus or dirt. One false step and it's easy to scrape away the covering and slide off the underlying ice.

Glacier tables are formed in a similar way. Individual rocks that land on the glacier will insulate the underlying ice. When the surrounding glacier ablates away, the rock remains perched on a pinnacle of ice looking much like a dining room table (see Fig. 4.6).

FIG. 4.6—Glacier tables are a type of debris cone. The insulated ice acts as a pedestal for the overlying rock.

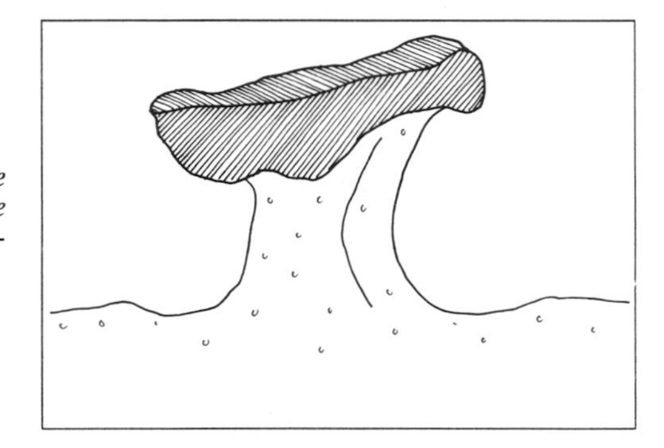

Sometimes rocks look like they have melted into the glacier surface instead of having been pedestaled up. Do not be fooled. This is a rock that has not yet reached the glacier surface. While buried, its dark color absorbs more heat than the surrounding glacier and causes the ice on top of it to melt first. When the glacier melts away to finally expose that buried rock, the thickness of the rock allows it to insulate the underlying ice, even though its upper surface still absorbs heat. Soon it too will become a glacier table.

Have a Cow

Ice that cracks off the end of a glacier terminating at the edge of a cliff or flowing into water is another form of ablation called calving. I have never actually seen a cow give birth, but when a glacier is calving it looks and sounds pretty rambunctious. There is a thunderous noise as the ice breaks off the glacier's face. Calving ice blocks from hanging glaciers become a powdery mass as the ice pulverizes on the rocks below. It continues thundering as the cloud rumbles down the mountain.

Calving from a tidewater glacier will cause a huge splash with water and ice bits flying all over, then a rumble and roar as the broken chunks roll and tumble, spreading waves in all directions. Indeed, I have heard that visitors to the Childs Glacier in Alaska wait along the shores of the Copper River for waves from the calving terminus to splash salmon onto the beach. (The best bait—a good book; see Fig. 4.7.)

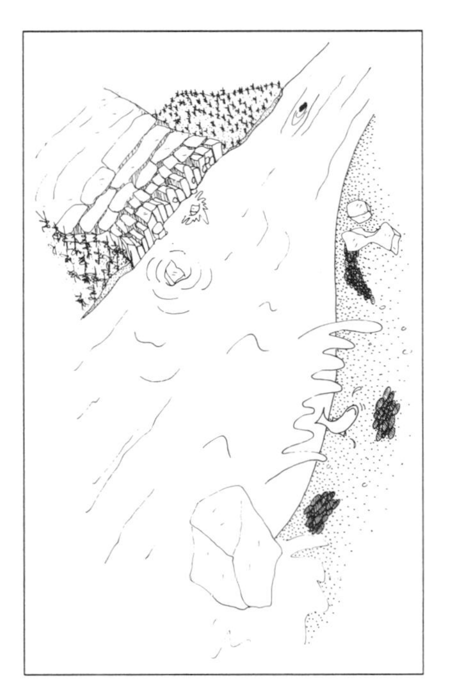

FIG. 4.7—Calving icebergs can create waves big enough to toss salmon onto the beach.

There are different names for icebergs according to where they come from and how big they are. For example, small icebergs that are about the size of a two-room cabin and float with less than 5 meters (16 ft) showing above sea level are called bergy bits. Growlers are even smaller, showing less than 1 meter (3 ft) above waterline.

While many icebergs are white with bubbly surface ice, those that come from deeper in the glacier appear dark blue with old, bubble-free ice. Those that look especially haggard, beaten up and black and blue, come from the glacier's underbelly, its sole. The blue is old, bubble-free ice and the black is rock debris picked up as the glacier slid over its bedrock bottom. The bottom bergs sit low in the water, weighted by the entrained rock debris.

The same terminology can be used to describe icebergs that calve into freshwater lakes and rivers. The only difference is that freshwater bergs float a little lower in the water because they do not have the buoyant help of denser seawater.

After floating for a while, small icebergs melt into a saddle-like shape. Melting actually occurs faster underwater, so after one end melts the berg flips over to allow the other end to sink and melt. The part in the middle is always in the water so it melts fastest, creating a two-humped iceberg.

Sometimes an iceberg will have a fluted or curtain-like surface. Although more common on bergs that calve into saltwater, fluted icebergs have been observed after calving into freshwater. The process that causes fluting is not completely understood. Somehow meltwater carves channels, either on the glacier front or within crevasses near the terminus.

The amount of wastage caused by calving is no small potatoes. A single, active tidewater glacier can shed nearly 22 billion kilograms (49 billion lb) of ice each day by calving icebergs into the sea. At an average density of 860 kilograms per cubic meter (55 lb/ft^3), the meltwater from that much ice could fill about eleven thousand swimming pools. Do I see smiling faces in the desert Southwest?

Tip the Scales

The difference between the amount of material that a glacier accumulates and the amount lost during ablation is called its mass balance. If a glacier gains more than it loses it will show a positive mass balance and be a healthy fat cat. A negative mass balance means that there is more expenditure than income, a situation that many of us, even glaciers, are getting used to these days. Although the large amount of reserve ice in a glacier can sustain this deficit spending for a while, there may come a time when it leaves town with no forwarding address. This was true for a small glacier on Mt. Timpanogos in Utah that clung to life for years before finally succumbing to starvation in 1987.

Calculating the mass balance of a glacier is not as easy as measuring the weight gain and loss of a dieting teenager. Beside the fact that it's not easy to heft a glacier onto a scale, measuring a glacier's income and expenditure is complicated by the complex patterns of accumulation and ablation, as well as its often tortuous shape (see Fig. 4.8).

For this reason there are only a half dozen or so glaciers in North America that have a long-term record of mass balance. Another dozen glaciers have short-term records performed in conjunction with a specific scientific study. Most of these guinea pig glaciers are off the beaten track so that visitors will not disturb the delicate instruments and precarious ablation stakes.

FIG. 4.8—Weighing a glacier is extremely difficult.

Zone of No Return

Another way to estimate the health of a temperate glacier is to compare the amount of surface area covered by snow with the total area of the glacier. This is called the accumulation area ratio (AAR) and is best observed on temperate glaciers when the period of melting has finished at the end of summer.

Most healthy mountain glaciers in North America have an AAR of about 60 percent. That is, 60 percent of the total surface area of the glacier accumulates more mass than it loses. Some glaciers can survive quite well with a 50 percent AAR because the mass that they do accumulate is abundant. This is especially true for glaciers along the Pacific coast, which are the first to collect snow from moisture-laden storms off the ocean.

There is no need to use this method for deciding the health of glacier remanié, rock glaciers, or debris-covered glaciers. They either receive their nourishment from some other source or are so cool under their space blanket of rock that it does not matter how much area is used for accumulating snow.

The minimum altitude of snow on the surface of a temperate glacier (not including small patches that may persist in shaded areas) at the end of an ablation season marks the boundary between the area of net accumulation and net ablation for which an AAR is calculated. This snow line is also the current equilibrium line (see Fig. 4.9).

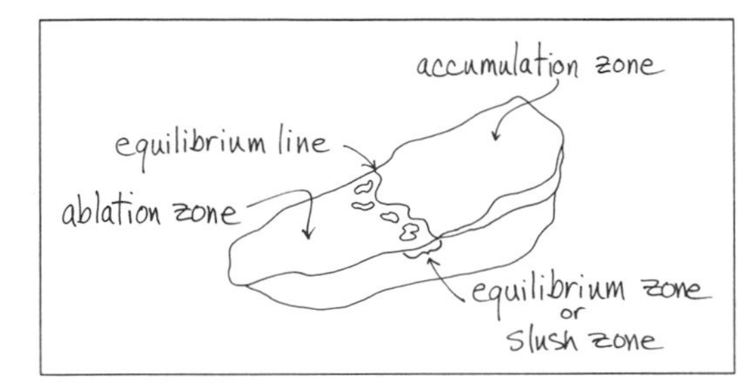

FIG. 4.9—Comparing a glacier's accumulation area with its total area gives an estimate of its health. The boundary between the accumulation zone and ablation zone is called the equilibrium line.

At the equilibrium line, the mass balance is zero. If you look down onto a glacier, the surface will look relatively clean and white where snow resides above the equilibrium line and rather dark with old firn or bluish with exposed ice below the annual equilibrium line. If the equilibrium line occurs in approximately the same place year after year, it will develop the banded appearance of sedimentary layers.

These exposed layers are sometimes called sedimentary ogives. The dust and debris collected on a glacier each ablation season provide a dark band of material. At the firn limit (where the firn has wasted away to expose bare ice) this dark band is visible on the glacier surface. The band will move down-glacier with the same speed as the surface velocity of the glacier, making room for another sedimentary band to develop above it (see Fig. 4.10a).

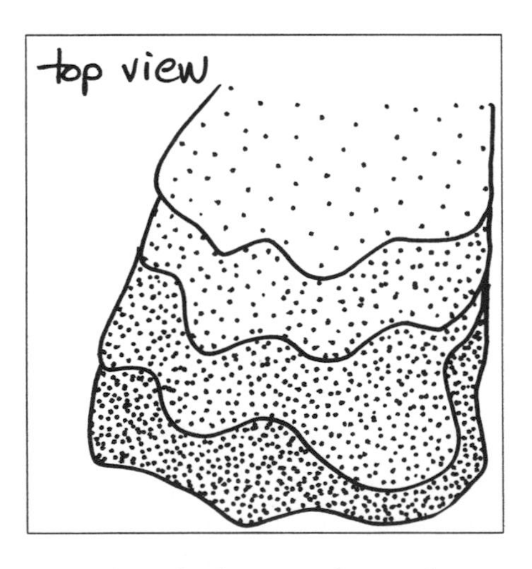

FIG 4.10a—If the equilibrium line occurs in approximately the same place every year, the associated firn limits can develop a banded appearance whose layers are sometimes called sedimentary ogives.

If the firn limit appears in approximately the same place each year, then a succession of dark to light bands will develop. The youngest firn layer will appear the lightest because it has accumulated only one year's worth of dirt. The older bands will be darker since they will have accumulated a layer of dirt each year (see Fig. 4.10b).

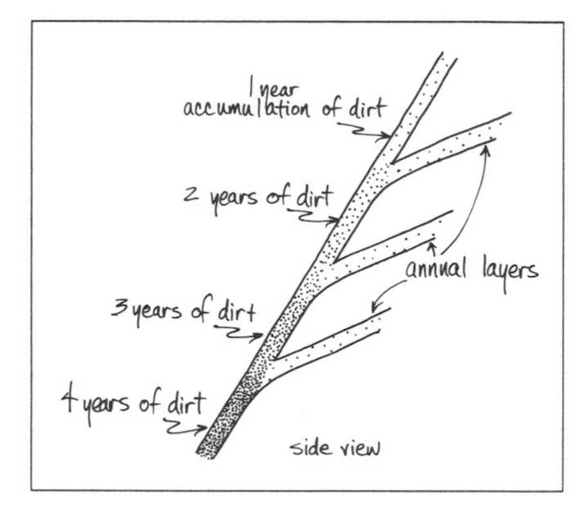

FIG. 4.10b—Each successive year of a banded firn limit will appear lighter. The youngest firn layer will appear the lightest because it has accumulated only one year's worth of dirt. The older bands appear darker since they have accumulated all the dirt from each succeeding layer.

On gently sloping or flat glaciers, the equilibrium line can be a zone of transition between relatively dry snow and bare ice. This area is usually messy with slushy snow and puddles. You will learn to avoid the slush zone if you spend much time travelling on glaciers. It can be very sloppy.

Sometimes the equilibrium line on a glacier is at an elevation above the highest extent of a glacier. If this is the case, you can bet that the glacier is struggling through an unhealthy season.

— Try This —

During the fall or toward the end of summer find a spot that overlooks the equilibrium line of a glacier. On a topographic map, draw the equilibrium line as you see it. Can you determine if the glacier's accumulation area is greater or less than 60 percent of the glacier's total area? Is the glacier healthy?

By returning to the same spot at the same time of year, you can develop a record of the equilibrium line altitude and thus keep track of the glacier's health.

This method of estimating a glacier's health is quite a bit easier than detailed mass-balance studies. It is so easy, in fact, that during late summer each year the United States Geological Survey conducts an annual checkup on about a hundred glaciers along the west coast of North America. In an aerial survey from Washington through British Columbia to Alaska, they observe the annual equilibrium line and calculate the accumulation area ratio. All of these examinations can be completed in just a few days of flight time.

Advance and Be Recognized

Many people equate a glacier's health to the relative position of its terminus. A glacier's terminus advances in a variety of ways that we'll learn about in Chapter Five. A glacier retreats backwards by ablating material at its snout.

Those glaciers that flow forward at a steady pace and move their terminus back by melting may indeed reflect some aspect of their health by the relative position of their snouts. A healthy glacier that accumulates a lot of material will build great bulk. As this material flows farther and farther downhill the terminus will advance to lower and lower elevations.

The snout of an advancing glacier will often have a bulbous shape with a steep face (see Fig. 4.11). Many splay crevasses may appear near the terminus of an advancing glacier.

FIG. 4.11—The snout of an advancing glacier is often bulbous with a steep face.

By contrast, a glacier that is losing material will have little excess to spread downhill. Its terminus will retreat to higher and higher elevations. It may even become stagnant, showing no forward movement at all. The snout of a retreating glacier will often have a gradually thinning terminus with a shallow sloping face (see Fig. 4.12).

FIG. 4.12—The snout of a retreating glacier usually has a shallow slope that gradually thins.

Sometimes a glacier can respond so quickly to changes in weather patterns that it will show a seasonal fluctuation in the position of its snout, advancing in the winter and retreating in the summer. Other glaciers respond more slowly, and it can be difficult to determine if they are responding to a current climate change or something that occurred during previous years. A complete study of a glacier's movement and its terrain characteristics is needed to determine how its length corresponds to changes in climate.

By contrast, a glacier that moves forward with sporadic bursts or ablates by calving will have little or no relation between its health and the position of its terminus. For example, some glaciers can experience a burst of speed that will advance their terminus without adding an appreciable amount of material. Their elongated shape is not an expression of improved health, just increased motion.

Because the rate of calving is related to the depth of water, tidewater glaciers are another example where the relative position of a terminus is no reflection of health. In fact, it appears that these glaciers advance and retreat in a cyclic manner that depends upon fjord geometry, having little to do with climate (see Figs. 4.13a, b, c).

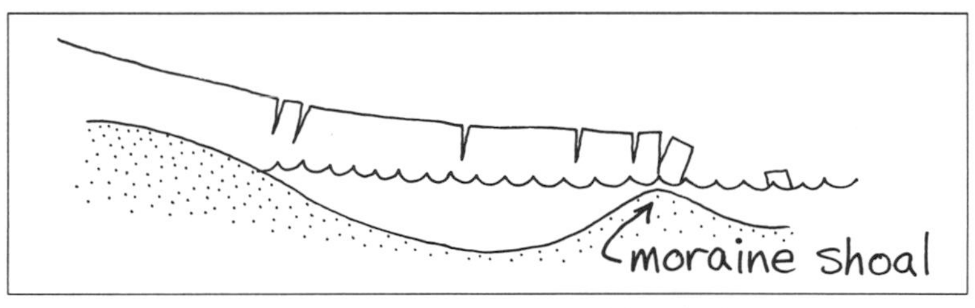

FIG. 4.13a—Tidewater glaciers that advance beyond their shallow moraine shoal will enter deep water and begin rapidly calving back.

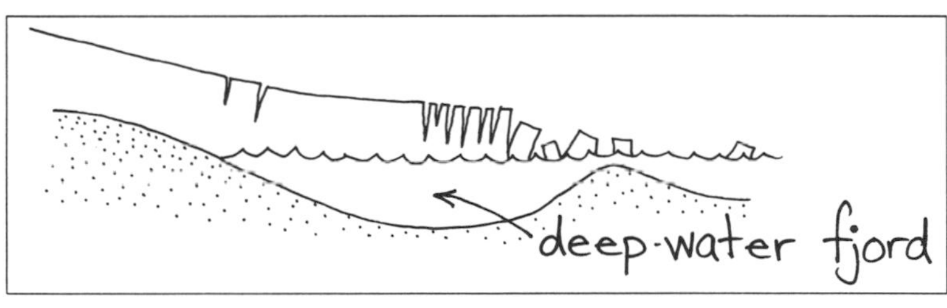

FIG. 4.13b—If a tidewater glacier retreats behind its shallow moraine shoal, the deep water of the fjord will cause it to retreat rapidly.

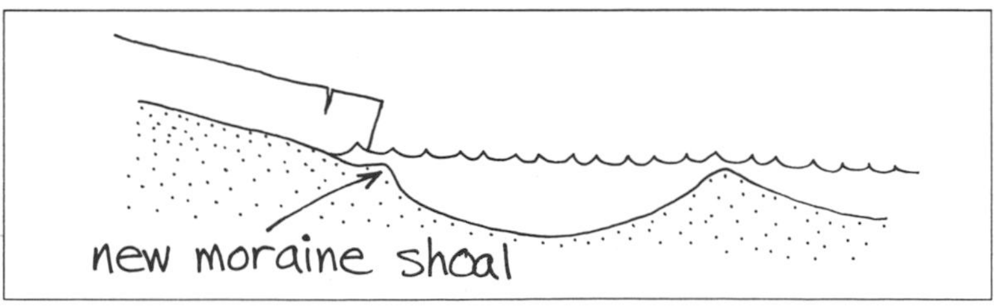

FIG. 4.13c—When a tidewater glacier once again advances, it does so slowly, as it builds a new shallow moraine shoal ahead of it.

When a tidewater glacier advances into a fjord, eroded rock material is pushed out from under its belly into a pile in front of it. This creates a moraine shoal that helps maintain shallow water. In shallow water the calving rate is slow. This keeps the glacier's forward motion faster than its backward calving. By pushing the shoal forward, the glacier can gradually advance quite nicely. In Alaska the character of some tidewater glaciers and their fjords has been studied to determine that it takes about a thousand years to push a moraine shoal forward and move the length of a fjord.

If a glacier tries to extend beyond its moraine shoal into deep ocean water, it will calve rapidly back to the shallow water of the shoal. If it likewise shifts behind the shoal, its terminus will again be in deep water of the gouged-out fjord. Retreat can be five to fifty times faster than the glacier's slow advance, calving back to its original shoreline within twenty to two hundred years.

An interesting place to view two similar glaciers in different terminal states is in the College Fjord of Alaska's Prince William Sound. Both glaciers have similar topography and similar climate. However, one is advancing and one is retreating. All of you Ivy Leaguers will be interested to know that the Harvard Glacier, at the head of the fjord, is the one slowly advancing, while Yale Glacier, adjacent to Harvard, is rapidly retreating.

Living in the Right Neighborhood

Clearly the inclement weather and climate patterns of mountainous terrain will influence how much snow and ice a glacier accumulates and loses over the course of time. Cold, wet winters will help nourish a glacier and cool, cloudy summers will help sustain its winter coat.

Snow can fall at any time of the year at the high elevations where glaciers are found. However, for most glaciers in North America there is a well-defined accumulation season when more material is gained than is lost and a corresponding ablation season when more material is lost than gained. The accumulation season roughly corresponds to winter and the ablation season roughly corresponds to summer.

Many glaciers in Alaska and northwestern Canada experience an accumulation season that is nearly ten months long, beginning in September and ending in late June. Several glaciers in the U.S. Rockies suffer short, less-than-six-month accumulation seasons lasting only from October to April.

Polar glaciers are part of the "economy class." Winter accumulation is usually very light, with snowfall averaging less than 25 centimeters (10 in) each year. However, the melt season is often very short or even nonexistent. Rock glaciers and debris-covered glaciers at these and lower latitudes are efficient in this way also. Often receiving no solid precipitation, these glaciers survive by being insulated from the summer sun.

"Gas guzzlers" are found in the coastal regions of North America from northern California to Alaska. Slopes facing the sea can accumulate 6 to 10 meters (20 to 30 ft) of snow each year and can lose as much or more during an average summer.

The health of a glacier obviously improves if the ablation season is short or if it is mostly cloudy with relatively cool temperatures. So next time you experience a drizzly, gray summer rejoice in the fact that your glaciers are happy and healthy.

In addition to the effects that weather and climate have on a glacier's health, many a glacier will succumb or succeed by living in the right neighborhood. This is especially true for small glaciers that have little variety in their prospects for nourishment.

Glaciers that are situated on the leeward side of a mountain are in a better position to survive than those on the windward side. Snow is scoured away from slopes facing the wind, inhibiting sufficient buildup for a proper glacier meal. The scoured snow will be redeposited on slopes away from the wind, helping to nourish the lee-slope glaciers. Similarly, those glaciers nestled at the base of a steep cliff can accumulate additional mass from snow and rock debris that avalanches and sloughs off the slopes above.

The sun's heat plays a major role in the amount of ablation a glacier experiences. Because the Earth's rotation axis is nearly perpendicular to the sun's rays, glaciers in the high-latitude Arctic have less heat to contend with than those in the low latitudes of California, Nevada, and Colorado (see Fig. 4.14).

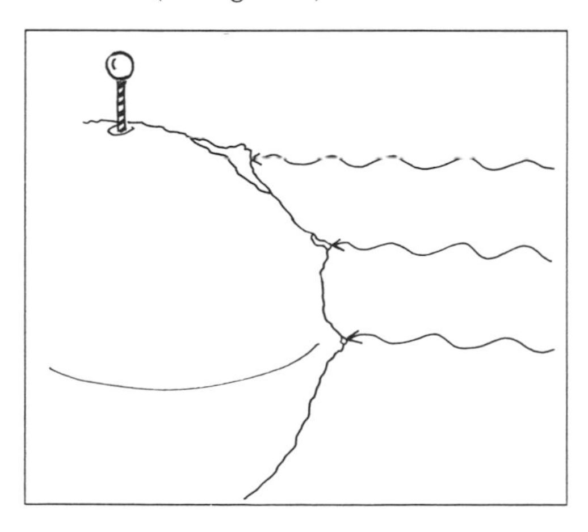

FIG. 4.14—The high angle of the sun at low latitudes in California and Colorado cause remaining glaciers to persist only at high elevations and mainly on shaded, north-facing slopes. The low angle of sunshine in the high latitudes of Alaska and the Yukon allow glaciers to exist at low elevations and even on south-facing slopes.

The high angle of the sun in the southwestern United States makes it additionally difficult for glaciers to escape the sun's heat. This is why you find most of California's remaining glaciers on the north side of mountains, nestled deep inside steep-walled, shaded cirques. On the other hand, the summer sun in northern Alaska and Canada swings around a broad arc of the horizon, warming glaciers on many aspects with nearly equal intensity. But the low angle of the sun and subsequent cool temperatures allow even small glaciers to survive on south slopes at these latitudes.

Elevation plays a major role in the location and health of glaciers. Obviously, a glacier at high elevation can remain cooler than one at low elevation. Usually this means that high elevation glaciers will thrive more easily than low elevation glaciers. However, in many places higher elevations can be drier than lower elevations, especially for those glaciers near the coast where precipitation is often confined to elevations below about 3,000 meters (10,000 ft) in a finitely thick layer of marine clouds. In this special case, the benefits of cooler temperatures may be offset by the hindrance of reduced nutrition.

You can see a glacier's reliance on climate and topography by observing the distribution of glaciers in North America. Glaciers in Alaska and northwestern Canada easily dip close to sea level. However, even the moist maritime climate of the Washington and southern British Columbia coast cannot offset the warmer latitude, and glaciers rarely extend below 1200 meters (4,000 ft) above sea level.

Farther inland at this latitude, toward Montana and Alberta, where winter accumulation is less than half of that along the coast, glaciers exist only above about 2200 meters (7,200 ft). Along America's sunbelt, glaciers must climb to higher and higher elevations to survive. Coastal glaciers in California exist only above about 3600 meters (11,800 ft) and the inland glaciers of Colorado are typically 3800 meters (12,500 ft) above sea level.

Medical Report

If you consider that most modern glaciers are remnants of the last major ice age ten to twenty thousand years ago, then it is reasonable to assume that their significantly shrunken stature points to a long period of decided ill health. Since that major ice age, however, there have been several periods of extended cooling.

The last cool period occurred during the mid-1800s to early 1900s. At this time glaciers were fattened up and stretched down-valley. Jalopies trundled to the snout of the Nisqually Glacier in Washington to retrieve ice for visitor cocktails at the Longmire Inn. Since then the Nisqually has retreated nearly 2 kilometers (1 mi) up-valley from the roadway.

The turn of the century also marked the heyday of the Glacier House on Roger's Pass in Canada. Because a restaurant car was too heavy to carry up the steep pass, the rail company built a well-supplied hotel and restaurant. Many Eastern families travelled to Glacier House and began some of the first glacier studies in North America on the nearby Illecillewaet Glacier. Now, only a hardy adventurer would attempt the stiff walk to the retreating glacier snout.

Alas, despite that brief period of good health, for the past sixty to one hundred years most glaciers in North America have been decreasing in size. This is primarily due to a warming climate. Many speculate that the warming trend is amplified by the human contribution of greenhouse gases. Others contend that the natural cycle of climatic temperatures will shift again to allow the approach of another ice age, despite what humans can do to bung up the process.

In either case, don't be discouraged. Even now there are new glaciers forming. Since the last eruption of Mt. Shasta in northern California, glaciers have become well established on its flanks. Although they are only about eight hundred years old and have not had enough time to gouge deep valleys, Shasta's glaciers are persisting well through this period of warming.

Likewise, nearly all of the volcanoes in North America have somewhat young glaciers, with real babies being born on Mt. St. Helens in Washington. The strongest evidence can be seen inside the crater where avalanching snow has piled deep at the base of the rim's steep cliffs. The snow is gradually turning to ice and may soon, perhaps before the turn of the next century, become an adult glacier.

CHAPTER
—5—

GALLOPING GLACIERS

Leaving my thick-soled "bunny" boots on, I stripped down to red long underwear, trying to remove anything metal that might interfere with a magnetic field. I then hefted a bulky magnetometer and strode about the glacier like one of those desperadoes you see in the park searching for lost diamond rings in the grass after an outdoor concert. I was looking for large bar magnets that had been buried in the glacier the year before.

Being so close to the magnetic north pole, we were having trouble because the energy field from our buried magnets was disrupted by occasional magnetic storms after a period of increased sunspot activity. Pretty wild stuff. The fluctuating magnetic field was so strongly apparent on the sensitive magnetometer that I had a vision of magnetic storms causing my new metal-edged skis to suddenly flip over, with me attached!

After letting my imagination subside, it turned out that I was able to find more of the magnets than others who tried. Some say it was because I had the least magnetic personality. However, I believe I was urged into the position of magnet finder because I wore the prettiest long underwear.

The located magnet positions were compared with where they had been placed the previous year to determine the glacier's annual velocity, its speed and direction of movement. These, and other measurements, were used to show that the glacier was moving about 30 centimeters (12 in) each day near its surface, gradually slowing at deeper and deeper depths, until it slid along its belly at a rate of about 10 centimeters (4 in) each day.

Compared to other mountain glaciers in North America (see Fig. 5.1), this one was travelling at or just above an average pace, moving ice on its surface about the length of a football field within a year. The range of speeds that make up this average pace is quite remarkable. For example, there are some glaciers that can gallop along the length of a football field each day! Still others, perhaps those that are semiretired, edge through the rest of their life at less than 1 centimeter (0.4 in) each year.

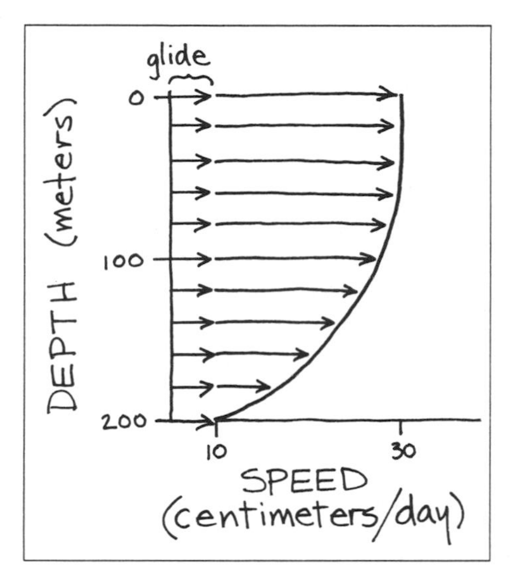

FIG.5.1—An average valley glacier in North America travels about 30 centimeters (12 in) each day near its surface and glides along its belly at about 10 centimeters (4 in) each day.

What forces a glacier to deform and move is simply its own weight. To this end, one eighteenth-century scientist compared glacier motion with softened wax. Next time you watch a dripping candle, imagine a large mountain with glaciers moving down its side.

Creepy Crawlies

One way that a glacier moves is through a method called creep, where ice deforms its internal structure. Within the snow and firn residing on top of a glacier, ice deforms by reorienting grains then bonding them together in a way that causes snow to densify and become glacier ice, as described in Chapter Three. This is a relatively fast motion that can carry particles several centimeters within a day or two.

Once the grains have developed into ice, they can further creep by slipping along internal planes of weakness within each ice crystal. The process is like spreading a deck of cards across the blackjack table. In this case, the cards would be stuck together with a very strong glue, and it would require a significantly strong and patient force to cause the glue to weaken and slide. As you can imagine, this slip motion is exceedingly slow and requires a great deal of force to initiate.

Because there is a large amount of weight exerted by a thick glacier, it can more easily deform its ice. It therefore moves faster than a thin, lighter-weight glacier, all other features being equal.

Along the same lines, the thick portion of a glacier can move faster than a thinner portion. For this reason, the accumulation zone near the head of a glacier usually moves faster than the thinner ablation zone. Sometimes a subtle change in climate can cause a fast-moving bulge of ice to develop in a glacier. These thick ice waves have been know to travel from head to toe two to six times faster than average.

Some visitor centers near glaciers in Alaska, Alberta, and Washington have time-lapse photographs of glaciers moving. Most films show icefalls, the steepest portions of glaciers, that typically move at a rate of about two and a half football fields each year.

The weight of a glacier and its resulting down-slope force is also affected by steepness. Steep terrain will cause more of the glacier's weight to be directed down the slope and less to press against the underlying bedrock, with less frictional resistance existing at the base of a steep glacier to hold it back. The result is that glacier ice moves slowly over flat terrain and speeds up over steep terrain, other things being equal (see Fig. 5.2).

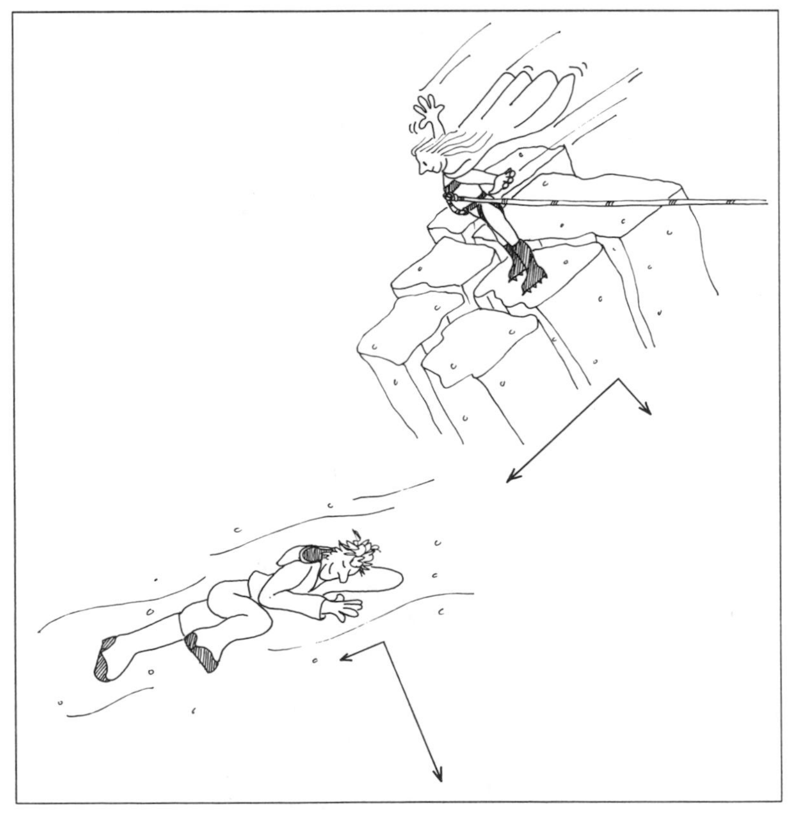

FIG. 5.2—Steep portions of glaciers travel faster than shallow portions because more of their weight is allowed to fall downhill and less is pressed against the slope.

— Try This —

Find a friendly glacier, one that is easy and safe to access and traverse. Collect a number of relatively large rocks from the surrounding moraine and place them in a straight line across the width of the glacier, 2 to 4 meters (7 to 13 ft) apart.

Come back in a month, or a year, and see what happened to your rocks. Providing that you found a relatively remote location, with no curious rock hounds, you ought to see the rocks form an arc shape across the glacier that illustrates the flow of surface ice.

Frictional resistance of surrounding bedrock at the glacier's edges and bottom cause it to slow down. Therefore, the fastest glacier motion, whether in an icefall or a gently sloping area, typically is found on the surface near the center of the glacier (see Fig. 5.3).

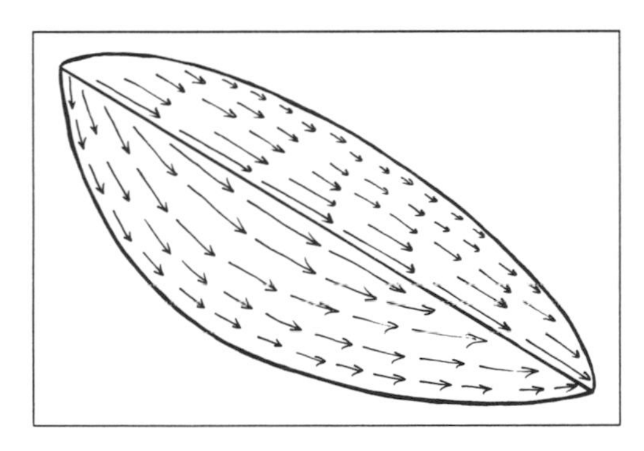

FIG. 5.3—The middle of a glacier usually travels faster than the edges because it is away from the frictional resistance of valley walls.

In addition to the little slip planes that exist inside the crystal fabric of ice, there are other planes of weakness that may exist between distinct layers of ice. This type of motion can provide substantial movement in a glacier. To find evidence of slip-plane motion, look for faults in the ice at a discontinuous interface between ice layers, especially near the terminus of a small, advancing glacier.

Glide

Although internal deformation within the ice can account for a large amount of motion, scientists have found that some glaciers move many times faster than is theoretically possible to move by creep. They have surmised that these fast-moving glaciers benefit from the added motion of gliding over the bedrock on their bellies. Cameras underneath glaciers, in ice caves and boreholes, have confirmed these notions.

Belly gliding is very slow in glaciers that are frozen to their bed surface. Glaciers that have abundant meltwater slide the fastest. There are several ways that sliding can happen, depending upon the amount of water available and the roughness of underlying bedrock.

Just about every glacier in North America generates some type of water near its sole because of heat coming up from the earth called geothermal heat, and frictional heating caused by ice movement. The film of water is nearly frictionless, allowing the glacier to slide over bedrock much more easily than if it were frozen to the underlying surface. It is similar for an ice skater gliding over a thin film of water created by frictional heating underneath the skate blades (see Fig. 5.4). Once the glacier is set into motion, it can create more meltwater by frictional heating as it slides past bedrock.

FIG. 5.4—Frictional heating can cause ice to melt. This type of meltwater helps ice skaters as well as glaciers to glide.

Regelation is a process that allows ice to move around small obstacles in the bedrock by first melting then refreezing. High pressures on ice will cause its melting temperature to decrease. Therefore, it will melt at temperatures colder than its normal melting point (0° C [32° F]). Low pressures will cause the melting temperature of ice to increase and meltwater will freeze at temperatures warmer than its normal freezing point (0° C [32° F]). Since ice on the uphill side of a small obstacle is under an extreme amount of pressure, it will melt. A very thin layer of water, only a few molecules thick, will then transfer the meltwater toward the lower pressure that exists on the downhill side of the obstacle. There it freezes. By transferring ice in this manner, from the front of an obstacle to the back, a glacier can slowly glide over rough bedrock.

— Try This —

To demonstrate the process of regelation, experiments have been designed using wire that is weighted to pull slowly through a block of ice. These experiments looped .45 millimeter (0.02 in) diameter wire around a sample of ice and weighted it with about 2 kilograms (4 lb).

At temperatures near -3.5° C (26° F) the wire took one hundred days to traverse through 1.3 centimeters (0.5 in) of ice. At -0.5° C (31° F) the wire traversed 35 centimeters (14 in) each day. Ice that was "cut" behind the wire healed and the ice block remained intact.

You can design an experiment like this in your freezer or the next time you visit a glacier. Remember, the warmer it is, the faster your experiment will proceed. However, at temperatures above 0° C (32° F), everything will melt before you can see any type of regelation.

Slip Sliding Away

Many times there are larger amounts of water available at the glacier's sole. This is especially true for temperate or subpolar glaciers that have a plumbing system connecting surface melt with the basal bedrock. The greater the amount of water available at the rock/ice interface, the less frictional resistance the glacier will feel as it slides over bedrock, and the faster it will move. For example, during late spring when glacier meltwater is abundant, velocities can be as much as fifteen to twenty times greater than wintertime movement on the same glacier.

Sometimes a mix-up in the plumbing system occurs. One part of a glacier may thicken abnormally while another part thins abnormally, causing up to 100 meters (300 ft) difference in some cases. After a substantial buildup of ice, the increased weight of the thickened portion will cause it to rapidly progress down-glacier. This may simultaneously relieve the plumbing system, spreading water out underneath the glacier to help float it downhill. The surfaces of floating glaciers have been observed to rise in response to an increased amount of belly water, confirming the idea that even a huge glacier can become buoyant.

Because the ice is lifted away from large areas of its rough bedrock surface, the glacier can shoot downhill at a very rapid rate, ten to a hundred times faster than its normal rate of motion (see Fig. 5.5). This surging behavior usually lasts about one year.

There are over two hundred North American glaciers that are known to surge. Like glacier outburst floods (discussed in Chapter

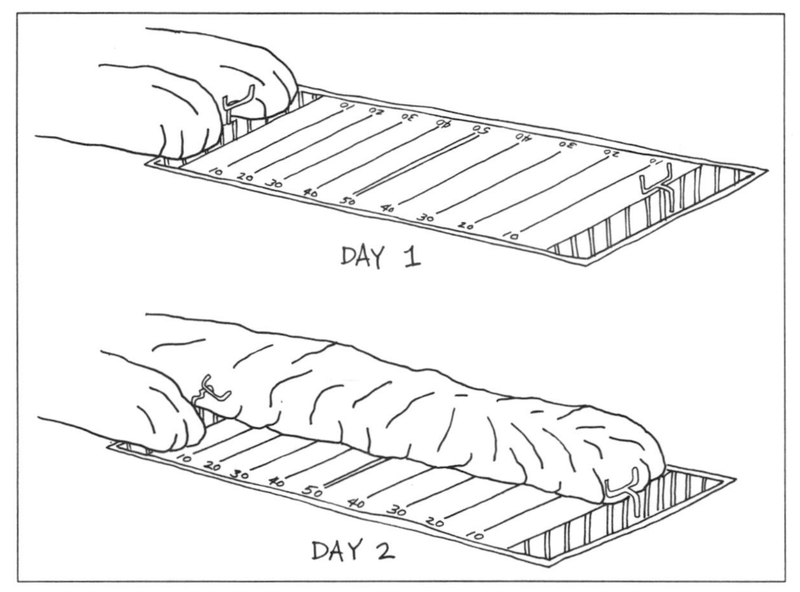

FIG. 5.5—When a glacier surges it can move ten to one hundred times faster than its normal rate of motion. This could cause a glacier, which normally moves the length of a football field in one year, to move that same distance in one day!

Seven), surge events are usually periodic, requiring fifteen to one hundred years to recharge between surges.

If a thickened portion of ice develops near the terminus, a subsequent downhill surge can extend its toe up to 10 kilometers (6 mi). Such an event surprised Mrs. H. E. Revell, who looked out her kitchen window one day in 1937 to see the Black Rapids Glacier barreling down on her. Fortunately the great wall of ice eventually stopped 240 meters (790 ft) short of the Black Rapids Road House along Alaska's Richardson Highway, which she operated with her husband.

Although most of North America's big surging glaciers are found in the St. Elias and Wrangell Mountains, they do not seem to depend upon any characteristic altitude, aspect, or bedrock type. Surging glaciers still threaten the Richardson Highway and also the Alaska Pipeline so it is important to know which glaciers surge, how often, and how far.

It is tough to be at the right place at the right time to see a glacier in the midst of its surge, but one day you may hear the rumbling sounds and observe the cracked and jumbled surface of a surging glacier. In Chapter Eight we'll learn some other clues to help you find glaciers that surge.

A Passage in Time

Knowing how fast a glacier travels ought to help you determine the time it would take for you to recover a toothbrush from the end of a glacier if it were dropped into a crevasse at the top. To help consider this problem, let me relay a little anecdote I heard recently.

In 1933 two park naturalists were walking toward the Lyell Glacier in Yosemite National Park. On the distant horizon they saw the regal silhouette of a longhorn mountain sheep. Both men were stunned. Sheep had been extinct in the park for fifty years.

They crept closer. The sheep remained still. Closer and closer the men ventured, and the sheep did not move. They came within 6 meters (20 ft), and the sheep stood fast. Then they saw it: The poor animal was stuck in the ice, dead. It was frozen in perfect condition and only just now thawing out of its icy tomb near the terminus of the Lyell Glacier.

Because it would take about two hundred and fifty years to travel the full length of the Lyell at that time, the men surmised that if the animal had fallen into a crevasse near the head of the glacier, it had been frozen for up to two hundred and fifty years before it melted out at the terminus. And since sheep had been extinct for fifty years, they narrowed the margin of entombment to between fifty and two hundred and fifty years.

Ice in most of the short valley glaciers and cirque glaciers in the coastal ranges of North America requires about a hundred years to travel from tip to toe. Where accumulation is less abundant and motion is slower, the same trip may take two hundred to two hundred and fifty years, perhaps like the Lyell Glacier in California.

Ice in larger glaciers of Alaska and northern Canada travel a full length in five hundred to a thousand years. Try tasting the difference between today's chemical-laden city water and that pure one-hundred- to a thousand-year-old water dripping off the snout of many North American glaciers.

CHAPTER
—6—

Creaking, Cracking, and Calving

I was daydreaming about snuggling into a featherbed when I dropped down the next crevasse. I detested the intrusion on my dreams and lamented the chore ahead to free myself. This time, however, I had dropped farther than before. I was truly stuck.

In a moment the sled we were pulling began to fall in on top of me. I yelled, but my voice was muffled by the loosened snow around me, then drowned out by the wind above. Although my skis and backpack were wedging me in so I could not fall any farther, a brief period of panic swept over me.

I was nearly upside down and a compass fell out of my pocket. I heard it clank for a long time as it descended into the crevasse that opened up wider below me. It was a long, long way to the bottom. I considered the option of removing my backpack to extricate myself but I needed its wedging support to prevent me from slipping farther.

My partner came back to free the sled and noticed I was underneath. Without a word, he pulled my rope into belay. I carefully released my pack, pressing myself against the crevasse walls to hold myself, and lifted it up to the surface. With my labored breath as the only sound, I gingerly released each ski and handed them out.

It was messy to clamber out of the slot because the new snow came pouring down over me with each start to a hand or foot hold. However, with the help of my belay, I was finally able to reach the surface where I silently hugged my rescuer and we exchanged a goggled glance of shared relief. It took just a few moments for me to reequip myself before we were off again to join the two other sled teams.

After several more hours of trudging we finally decided to bivouac and wait out the storm. We were all tired, crevasses were increasing in size and numbers, the visibility was next to nothing, and darkness was imminent.

Quakes in the Night

I spent a fitful night as the storm raged around my wee tent. Every once in a while an ice quake would jar me alert. We were on a glacier in the beginning stages of a surge and it was pulling apart. These quakes were just the first indication that a fissure was opening in the stretching ice. Unlike an earthquake, where a low rumbling noise can be heard, the fractures that cause ice quakes make a hissing sound as they zip open. These fractures were the first stage to new crevasses. I just prayed the ice did not open up fast and swallow me whole!

As I lay in my sleeping bag, shivering with those eerie thoughts, I remembered a job I had just before Mt. St. Helens erupted in 1980. I was in charge of counting her quakes. It was sometimes difficult to distinguish between earthquakes and ice quakes because there were so many of both. Magma flowing under St. Helens not only caused rocks to fracture but also generated "hot" spots under her glaciers. It is believed that the hot spots caused increased melting, and this in turn lubricated the glaciers' bellies so they could slide faster down the mountain. The increased speed caused the ice to crack, forming crevasses and accompanying ice quakes.

— *Try This* —

Bring out some of your favorite Silly Putty®. Roll it into a cigar shape then grasp each end and pull the putty slowly apart. It droops and sags but essentially stays together as one piece.

Now ball it up and roll it back into a cigar. This time, pull it apart rapidly and see how it breaks, or fractures apart. Ice is like Silly Putty®. When moving slowly it flows and deforms. When moving rapidly it fractures apart.

Fortunately for me, my tent was not in a place where the ice would open, dramatically swallow me, then clap shut like some horror story. (If I had believed more strongly in my scientific principles that night, I probably would have been able to sleep.) Actually, the first jolting crack of glacier ice usually causes only a hairline fracture that may or may not be visible on the ice surface. It will take several hours or days, sometimes weeks, to open large enough for a person to slip through. When a fracture in ice opens and its surfaces are separated, it is called a crevasse. If the fracture surfaces remain touching, they may heal themselves and never open at all.

Open Fissures

The width, depth, and breadth of a crevasse depends on how fast and how much of the ice is moving at any one time. Near icefalls, where glacier movement is rapid, a crevasse can open as much as 50 centimeters (20 in) a day. Icefalls are probably not good places to stretch out your bedroll. In a slower part of a glacier, crevasses can take nearly a month to open as wide as 50 centimeters (20 in).

Fractures occur when the ice encounters a force greater than it can bear. And even though my falling compass had me believing that I had a direct route to China, crevasses have a finite depth. We can estimate how deep they become by considering the variations in strength and stress between the glacier surface and its bedrock base.

Near the bottom of a glacier the ice is strong with large, well-bonded crystals. The weight of overlying ice at this depth is large and it squeezes the ice further. When stresses try to pull this ice apart, it simply flows and deforms like thick, gooey honey.

Near the surface of a glacier where ice is just beginning to form, it is full of tiny flaws and weakly bonded crystals. When this ice is pulled, there is little overburden pressure to help squeeze and deform the crystals. It breaks apart in a brittle manner, like glass.

Ice fractures that begin at the surface of a glacier penetrate to the deep ice layers, then stop. Theoretical models of the stress and movement in temperate glaciers indicate that this depth is 25 to 30 meters (80 to 100 ft) about the same depth as a seven-story building is tall (see Fig. 6.1).

FIG. 6.1—The depth of a crevasse is usually about 25 to 30 meters (80 to 100 ft), breaking open in the brittle upper portion of the glacier and squeezing shut in the deep, malleable ice.

— *Try This* —

Roll your Silly Putty® back up into a cigar shape and put it in the freezer for an hour or two. After it cools it becomes more brittle. Pull it apart now and see how easy it is to break.

Ice is more brittle to break when it is cold. For this reason crevasses in polar or subpolar glaciers can be deeper than those in temperate glaciers. Sometimes portions of glaciers in the cold, high elevations will have deeper crevasses than those in the warmer, lower elevations.

Can't Cross Here

Crevasses occur where ice is moving at a speed that is different from its surroundings. The crevasses described above are transverse crevasses because they form in areas of extending flow. That is, the downhill ice is moving away from the uphill ice. Extending flow is common in the accumulation zones of glaciers and at the upper edges of steep sections (see Fig. 6.2).

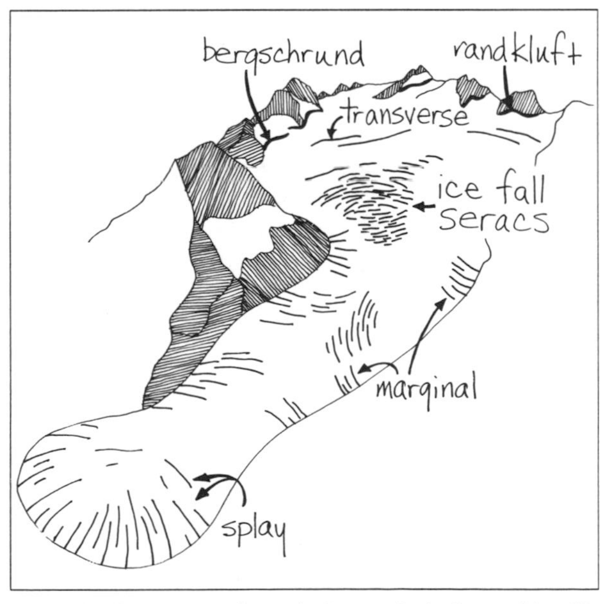

FIG. 6.2—Crevasses are located wherever the ice is stretching, like at the top part of steep sections, in the accumulation area, near the sides, and where it is spreading out at the terminus.

Where a glacier is compressing no cracks can form because there is no mechanism to pull the ice apart. These areas are commonly below a steeper section where the ice is slowing down.

When the ice moves very fast, such as within glaciers that are surging or over steep sections of icefalls, seracs may form. These isolated columns of ice are the result of extensive fracturing (see Fig. 6.3).

FIG. 6.3—The isolated ice columns of seracs form when crevassing is extensive, such as over an icefall.

Bergschrunds are crevasses that develop at the head of glaciers. These fissures separate flowing ice from a relatively stagnant ice apron that develops on steep rock faces above many mountain glaciers. This apron of ice does not move very much because there is usually not enough mass for it to flow under its own weight. Wind probably scours much of the winter's snow away and prevents deep layers of ice from forming. If the ice apron is missing, the gap between a moving glacier and its headwall of rock is called a randkluft.

Near the side of valley glaciers, marginal crevasses form. Ice is slowed quite a bit at the glacier's edge as it tries to slide past the stationary bedrock. Meanwhile ice in the middle of the glacier runs merrily along at its normal speed. As the middle ice shears past the slower edge ice there is a shearing force created that is oriented in the same direction as the flowing ice.

The force that actually pulls ice apart to form crevasses is a component of this shearing force. It is called a tensile force. It is oriented about 45 degrees from the direction of the shear force and causes crevasses to break open at a 45-degree angle from the glacier's edge (see Fig. 6.4). In an area of extending flow marginal crevasses may cut all the way across a glacier and be difficult to distinguish from transverse crevasses. In an area of compressing flow side crevasses only extend part way toward the middle of the glacier then pinch closed.

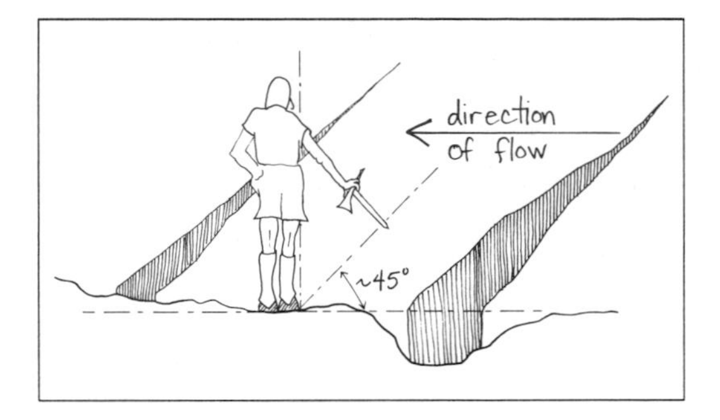

FIG. 6.4—Sometimes you can enter a region of marginal crevasses by heading about 45 degrees up-glacier to follow the same direction in which the slots open.

Sometimes the toe of a glacier will be away from its confining valley walls. When this occurs, the flowing ice spreads out in all directions. These spreading lobes form a splay pattern of crevasses that can be very difficult to negotiate.

— *TRY THIS* —

Next time you come to the edge of a glacier notice how difficult it may be to find a direct route straight across the glacier—especially if there are many crevasses. Now face uphill and look along the edge of the valley wall. Take out your compass and turn yourself to face 45 degrees away from the edge. Can you walk safely between crevasses if you stepped onto the ice in this direction?

This may not always work because glaciers and their carved valleys come in irregular shapes. Also, old side crevasses turn down-glacier as they close and new stresses are imposed.

HIDE AND SEEK

Even if you know where crevasses might form, they are often hidden under a blanket of snow. Many times this snow covering is very weak, especially in fall and early winter when the snow cover is still relatively thin. Later in winter sufficient snow may accumulate to develop stronger bridges across the buried slots. However, even at this time of year, the snow covering may become brittle enough to break away under the added weight of an unsuspecting glacier explorer.

During spring, the snow begins to melt away or break open and reveal most of the crevasses. However, even though a crevasse may be completely open and visible, snow can drape over its lip and hide the true edge. Those pieces of snow that survive the spring breakup may provide a bridge across a crevasse opening if sufficiently strong.

Visible or not, it helps to know where crevasses are most likely to be found. Hopefully, enough information was presented here to help you figure that out. In Chapter Nine we discuss how to safely negotiate crevassed terrain and what to do if you survive a fall into a crevasse.

Look Out, It's Falling

Crevasses that develop in icefalls or near the toe of a hanging glacier form seracs that can cause ice avalanches when they fall. The timing and extent of ice avalanches intrigue many scientists and is of great concern for those who travel on glaciers.

There is some correlation between time of year and ice avalanches on the warm, low-elevation portion of glaciers. During the peak melting season of late summer and early fall, increased water at the bed surface causes increased ice movement. This, in turn, allows more ice to fall off precipices and form avalanches. During other times of the year the ice is frozen to the bedrock and cannot move as easily. High-elevation glaciers are usually frozen to the bedrock and have no seasonal cycle of avalanche activity.

Normal glacier movement can cause ice avalanching at any time and at any elevation (see Fig. 6.5). A complicated system of mass buildup and discharge within the glacier ice enhances differential motion at icefalls, producing cyclic or periodic avalanche occurrences. Only when a glacier is closely monitored can this type of motion and subsequent avalanching be predicted. Just a handful of glaciers in North America have been instrumented like this and no information is available on a continual basis to help climbers. In addition, extra forces of earthquakes and volcanic tremors can trigger ice avalanching at unpredictable times.

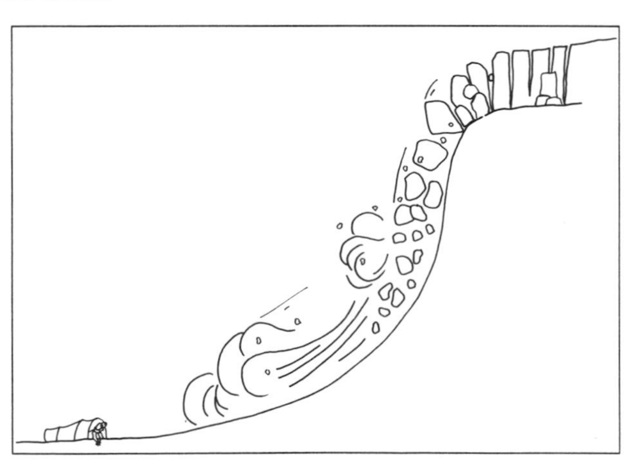

FIG. 6.5—Ice avalanches can happen at any time of year and at any time of day or night.

Some observers suggest that ice avalanches are more frequent during the warm part of the day, when melting is maximum. Perhaps meltwater can reach the base of exposed seracs, but this would occur only in very thin, temperate glaciers or serac fields that are open to the basal rock. Also, no daily afternoon pattern has been recorded within the resolution of scientific methods developed so far. Indeed, one scientist has shown that the strongest peak in ice avalanche activity is during the early morning hours when the ice is most brittle.

A more plausible theory suggests that loose snow slides could trigger a precarious serac to fall. The snow slides could initiate during afternoon warming.

Recent studies on the run-out distance of ice avalanches have shown that the steeper the slope over which an avalanche falls, the more turbulent friction it will have. This causes it to travel less far than one of the same mass and velocity travelling over a shallower slope. However, I doubt you would have enough time to gather all the numbers needed to plug into an equation that would determine if you need to get out of the way while an avalanche is racing toward you.

— *Try This* —

If you find yourself near an active ice avalanche area, take your watch out and time each event. (You may have to use a calendar to time the slow-cycle events.) If you cannot sit idly for that long, look for evidence of avalanching every time you pass by the suspected ice avalanche area. You should see ice debris on top of the normally smooth glacier surface. Can you see any correlation in avalanche activity with time of day? With time of year?

Deep-water Trouble

Calving glaciers are related to ice avalanches. The difference is that when a chunk of ice breaks off the end of a glacier it falls into a body of water rather than tumbling down a mountainside. As mentioned in Chapter Four, the rate of calving seems to be dependent upon the depth of water. In deep water the floating ice has plenty of room to bend and crack. Therefore, a large amount of calving can occur. In shallow water, the ice may not be able to bend far enough to crack. It may even be grounded, and with only the forces associated with flowing ice its calving rate is usually slow.

When the calving occurs in tidal water, there is some suggestion that tidal forces trigger calving events. As an idle youth, I spent some time watching the Hubbard Glacier in southeast Alaska calve into the Russell Fjord. I wanted to climb onto the snout but was afraid that the ice I chose to climb would drop into the sea. Calving events occurred at least every five minutes. I was not watching long enough to determine if there was a correlation with high tide, but I abandoned my desire to be on the glacier anyway.

The Lower 48 has no glaciers that end in the sea. However, there are a few that end in lake water. For example, the South Cascade Glacier in the northern Washington Cascades frequently calves into a small lake at its terminus. These splashy events have played havoc with nearby gaging stations on the South Fork of the Cascade River. Look for icebergs in the next mountain lake you visit. There may be a calving glacier nearby.

Thrusts and Moats

Sometimes fractures develop on a glacier that are not associated with crevassing. For example, thrust faults can form between two distinct layers of ice. This is especially true at the boundary between two annual layers where dust and dirt create a weakened interface. Thrust faults are most commonly seen near the toe of small glaciers whose annual layers have maintained their integrity.

Thrust faults may cause a layer of ice to push forward over another and form an overhang (see Fig. 6.6). Another way to recognize a thrust fault is to look for some feature perpendicular to the fault (like an old drain channel) that has been displaced.

FIG. 6.6—Different layers of ice can be pushed over one another by thrust faults.

Moats are another type of fissure unrelated to crevassing (see Fig. 6.7). These are openings near the sides of a shrinking glacier that are caused by melting. Because surrounding rocks are a darker color than ice, they absorb more heat. This enhances melting at glacier margins. Like crevasses, moats can be hidden by a blanket of snow. This creates a formidable challenge for those entering and leaving glaciers.

FIG. 6.7—Ice often melts away when it is near dark-colored rock. This creates moats near the edges of glaciers.

CHAPTER
—7—

Things That Go Spurt

I held tightly onto Daddy's hand and watched little goose bumps spread over my arms. It was cold. I was afraid of slipping on wet rocks and falling into one of the many icy streams. We were walking inside the stomach of a glacier—a huge cavern with eerie blue walls drip, drip, dripping with water. I was a little girl in Paradise.

These were the Paradise Ice Caves, just a short walk from the Paradise visitor center in Mt. Rainier National Park. It was 1959 and although the caves were in a remnant portion of the Paradise Glacier, sometimes referred to as the Stevens Lobe, they were no less impressive.

I have since been back to the caves. The retreating glacier has reduced their grandeur, but it is still possible to walk inside and see the exposed ice that was formed over a hundred years ago. Some of the ice is so clear it is like peering deep into the glacier's soul through a large plate-glass window. In other places bubbles of air that my grandfather used to breath are pressed and squeezed to such an extent that it looks as if they will burst out at any moment.

The Paradise Ice Caves are only a couple of many others in North America that are available for exploration by the adventurous spelunker. Glacier caves are created by draining meltwater that sculpts channels and conduits through and under the glacier. Caves are often enlarged at the edge of a glacier as warm air is sucked underneath the ice, enhancing melting.

The best place to look for cave entrances is at the glacier terminus where meltwater streams exit from underneath the glacier. The best time to go spelunking is in the fall when summer melt has slowed and winter snow has not yet covered the entrance. Because moving ice can deform and close openings, the largest caves will be found under relatively stagnant or retreating glaciers.

I do not usually enjoy spelunking. So many caves begin with a tiny crawl space before opening up into larger caverns. The thought of being squashed between the fragile roof of ice and a cold-water stream is not a pleasant one for me. In fact, the last cave that I contemplated exploring had a dead mountain goat guarding the entrance. Apparently the poor fellow had fallen through thin ice over the cave entrance and perhaps broken a leg when he landed on the streambed rocks. It reminded me of a dance performance I saw during a recent visit to New York, "Dead Goats Don't Faint." I hoped that the goat had had a chance to faint before it succumbed to the icy waters of the glacier.

Runnels and Rivulets

Just how does all this water reach the glacier's belly to form ice caves? It is not a straight path, believe me.

If you have ever been fascinated by the undulating pattern on the surface of melting spring snow, you will have an idea about how surface meltwater begins its trip to the bottom. Free water on a glacier (that is, melted out and not locked into an ice crystal) begins to percolate through the surface snowpack by flowing closely around each grain in its path. This is a long and circuitous trail. However, at many places within the snow enhanced melting allows small fingers of flow to penetrate to deeper depths (see Fig. 7.1). Once a flow finger develops, it creates an easy pathway for further meltwater. Soon the finger will extend to a full-fledged drain channel, allowing meltwater to drain all the way through the snowpack.

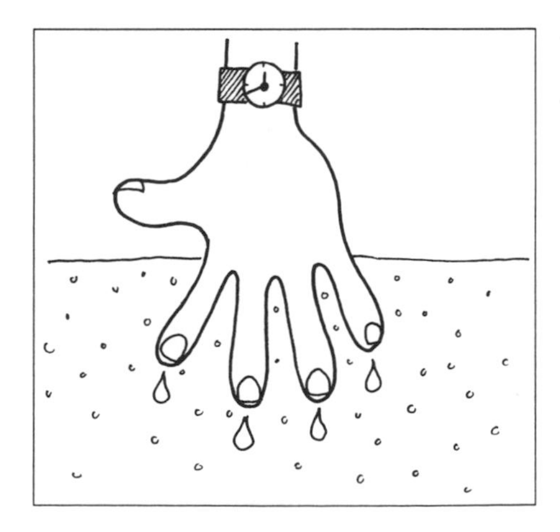

FIG. 7.1—Water begins to percolate through snow in small channels, or flow fingers, where enhanced melting has created a pathway.

These drain channels, which are full of wet, slushy snow, compress more rapidly than the surrounding drier snow. This causes the surface to develop dimples where the flow fingers exist. If the drain channels occur on a sloping surface, the dimples look more like runnels.

On a typical spring snow cover the drain channels are spaced about .5 meter (2 ft) apart. They can form beautiful patterns on the glacier surface that show how meltwater has drained downhill through the snow and young firn. Be aware, when the dimples and runnels freeze at night, glissaders and skiers will find themselves in for a hard, bumpy ride!

Halting Ice

Meltwater on the surface of a polar glacier rarely penetrates very deep. Most often it will reach cold snow and refreeze before it can drain farther. However, water draining through the snow and firn cover of a temperate glacier is only the first stage of an amazing journey.

Because snow and firn covering a temperate glacier are already very close to their melting temperature, water can drain completely through this covering. But when the free water reaches underlying layers of glacier ice, it is discouraged from penetrating further.

Certainly if ice is impermeable to air (as you found out in Chapter Three with the blue-lip test), it must be impermeable to water. For the most part this is true. In fact, if you were to poke a hole through the névé of a temperate glacier, you would eventually locate a water table, a level below the glacier surface that is saturated with water. The water simply pools above the ice and floods the overlying firn. This water table is a real menace for those of us who drill holes into glaciers. More than once I have gotten my auger stuck from liquid suction in saturated firn. I have developed unusually broad shoulders trying to retrieve these bothersome instruments.

On the ice surface water drains downhill by eroding channels and developing streams. These rivulets are most visible in the ablation zone of a glacier where the ice surface is exposed. It is usually excellent drinking water because there are few animals on a glacier that can spread harmful parasites such as *Giardia*, and most of today's glaciers are well away from pollution centers.

Mill Streams

Although many streams carry surface water off to the sides of a glacier, some flow into crevasses. Here, the flowing water either connects with some old buried stream, or penetrates through the bottom of the crevasse and develops a deep conduit of its own.

When the crevasse closes, heat carried by the flowing meltwater maintains a pothole into the glacier. This is called a moulin or glacier mill. Many moulins are as deep or slightly deeper than the original crevasse, about 30 to 40 meters (100 to 130 ft). However, some have been surveyed at over 100 meters (300 ft) deep (see Fig. 7.2).

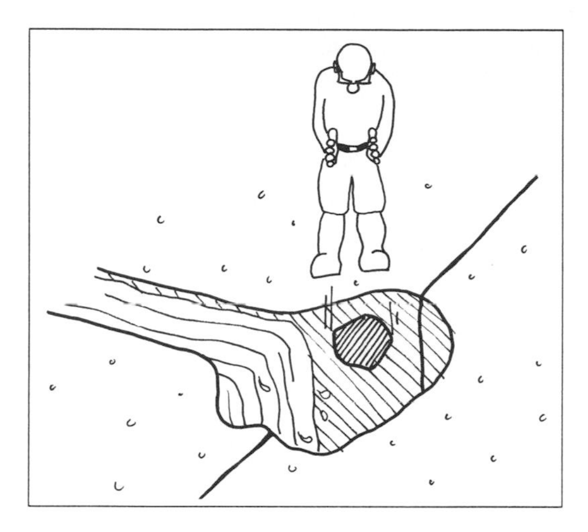

FIG. 7.2—You can check the depth of a moulin by dropping a rock down its meltwater channel and listening for it to reach bottom.

The best place to find moulins is near the centerline in the ablation zone of a glacier. Ideally the area up-glacier from a moulin or group of moulins will be flat or gently sloping so that plenty of meltwater flows on the glacier surface. Moulins develop when a crevasse opens up and intersects a stream; this could typically occur within a region of transverse crevasses. (Marginal crevasses may allow meltwater to find an easier path to the edge of a glacier and would be less likely to develop moulins.)

As a moulin is carried down with the moving glacier, new crevasses open upstream and create new moulins. For this reason it is not uncommon to find several moulins within a small region of the glacier. Moulins down-glacier will be old and perhaps dry, while moulins up-glacier will be new and roaring with flowing water.

— Try This —

Next time you happen upon a moulin, pick up a rock and drop it down the hole. The echoing sound of the rock as it descends will give you a pretty good idea about the depth of this hole. Some of the more impressive drops will take over four seconds. Be careful not to fall in.

It is unclear just how water from a moulin reaches the bottom of a glacier. For some temperate glaciers the surface meltwater requires just a few hours to reach the bottom. On other glaciers, water takes several days to wend its way through an intricate path of channels and chambers. In more fantastic circumstances, water is held for several years in a glacier before it finally escapes, often catastrophically.

Even without surface melting, there can be a thin layer of belly water under a glacier. Some of this water may be produced by frictional heating and melting of the ice as it flows over bedrock. Although more common in temperate glaciers, fast-moving polar glaciers can cause melt in this way as well. Other meltwater may be produced by geothermal heating from below.

Pooh Sticks

Meltwater coming from a glacier is an important factor in many of our lives. It is used for irrigating agricultural lands, for drinking, and for hydroelectric power. Glacier meltwater can also control the speed of a glacier and can create some imposing hazards. This is why the plumbing system of a glacier is studied so intently by many scientists.

During winter, water available to flow from a glacier is usually a small amount produced by frictional or geothermal heating at the glacier's underbelly. However, during the summer, at the height of melt season, the outflow rivers of most temperate glaciers will involve over ten times as much water as their winter flow.

In addition to the increased volume, summer outflow of temperate glaciers also shows a large fluctuation at different times during the day. Maximum summer outflow usually occurs just a few hours after the maximum period of melt; this would be around dinner time, say 5 or 6 P.M. Minimum outflow occurs during the coldest period, late night and early morning. The amount of water pouring out of the glacier in the late afternoon can be over two times as much as trickles out in the early morning. On the other hand, during the winter season, without variations in daytime heating and subsequent surface melting, the daily flow of water from a glacier is nearly constant.

Average summer outflow has been measured for small, temperate valley glaciers at a few thousand liters each second. At this rate it would take less than ten minutes to fill an Olympic-sized swimming pool. Larger glaciers, like those in Canada and Alaska, which have more material to melt, can fill the same size pool in much less than a minute.

Dam This Traffic Jam

Not all the water that flows through and around a glacier flows freely. Many things can happen with this intricate plumbing system.

— Try This —

Play "pooh sticks" in a glacial stream. On your way to visit a glacier in the early morning, look for a stream coming out of its terminus. It will help if you can locate a stream that is in a single channel, not braided into several small channels.

Designate two points to mark the beginning and ending of a race course in the stream. The longer the distance is, the easier it will be for you to estimate flow, so you may want to work with a partner.

Drop a wooden stick into the stream at your first marker. Count the number of seconds it takes for the stick to float to your second marker. Remember that time.

On your way back in the afternoon, stop at the outflow stream again. Drop another stick into the stream and count the time it takes for this one to flow the same distance as your morning stick. It will probably take less time.

If you could measure the cross-sectional area of the stream during its minimum flow and during its maximum flow, then you could calculate the volume rate of water emanating from the glacier. Do you think there is enough water to power a small hydro plant? To water a row of houses in Beverly Hills? To provide showers for athletes at the Olympics?

All of us who have had experience with backed-up sinks and toilets certainly can relate to this problem.

Water can dam up in and around glaciers for a variety of reasons. Sometimes the water actually is diverted back up to the glacier surface. This causes a "negative mill" or geyser to form. These gurgly displays are rare so be sure to ogle and gawk if you ever find one.

Often the movement of the glacier will close small conduits. This is especially true after a conduit loses its "roto-rooter" (freely flowing water whose heat and pressure can keep conduits open). If the little conduits offer the only exit for internal cavities or chambers, then their blockage will cause these voids to eventually fill with water.

A water-filled cavity may exist for long periods of time or may find a new exit for drainage. A small flood often develops when the cavity eventually drains. Floods from single draining cavities or several linked cavities have been known to cause outflow streams from the Athabasca Glacier in Canada to soar to a level well above average and remain so for several days.

Glacial movement can also cause internal cavities and chambers near the base of the glacier to alternately open and close. These types of chambers are common under glaciers that reside on steep, undulating terrain, like that of a stratified volcano (common along the coast ranges of North America). On the down-slope side of a break in the slope, the ice will leave contact with the bed surface for a while. It is almost the same trajectory that a downhill skier makes when he or she "catches air" over a rolling knoll.

The glacier ice reattaches to the bedrock at the leading edge of the cavity with enough squeezing pressure to prevent the passage of any water (see Fig. 7.3). The cavity will fill and additional water pressure may allow it to enlarge. Eventually the cavity becomes big enough to connect with the main plumbing system, at which time it floods out.

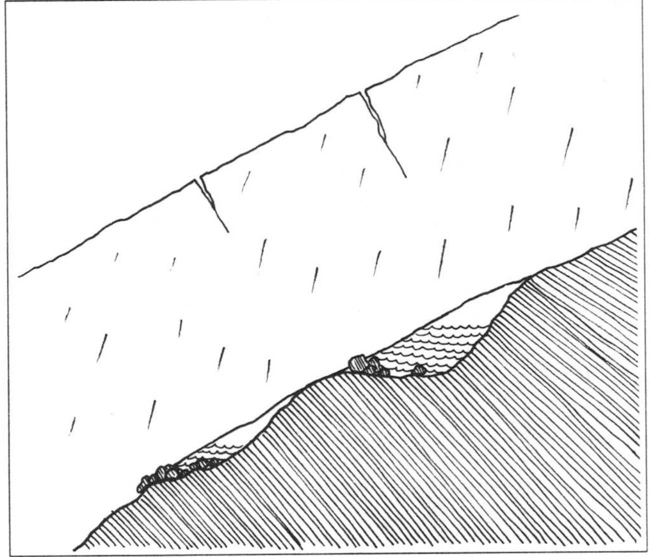

FIG. 7.3—Sometimes meltwater is trapped underneath a glacier. This is common on stratovolcanoes that have undulating layers of bedrock. When finally released, the outburst flow can cause catastrophic floods.

These types of outburst floods can carry a hundred to a thousand times more water than the mean daily flow. Travelling at about 20 kilometers per hour (10 mph), lasting several hours, and sounding like a freight train as they entrain soil, rocks, and trees, the floods present a significant hazard to any object or person in their way. They are most common during the peak melt periods—late summer or fall in the late afternoon or early evening.

Significantly larger pools of dammed water can occur at the margins of glaciers where streams and rivers are dammed by the moving tongue of ice (see Fig. 7.4). This causes a lake to form behind the dammed stream. Eventually the buildup of water pressure in the lake becomes great enough to actually lift the glacier up and out of the way so water can drain out underneath the glacier—sort of a lift, separate, and swoosh mechanism.

The sudden draining of a dammed lake causes outburst floods that are so great they have been given their own word, jökulhlaup. The Knik Glacier used to dam the outflow of Lake George (northeast of Anchorage), causing an annual jökulhlaup to release over 700 billion liters (200 billion gal) of water in less than a day.

Jökulhlaup is an Icelandic term for glacier outburst floods. Although you might try pronouncing it "yo-kul-h-loip," I have never been able to say it properly enough for an Icelander to understand me.

Most jökulhlaups occur at regular intervals. After a lake drains it will take only a little while for the moving glacier to dam its outflow again. Then the lake will recharge. Some jökulhlaups occur annually, others require several years to build up before each discharge. The jökulhlaup from Lake George occurred so regularly that the Alaska railroad included costs to repair damages from the expected floods in each annual budget. (No floods have occurred since 1966 because the Knik has been retreating.)

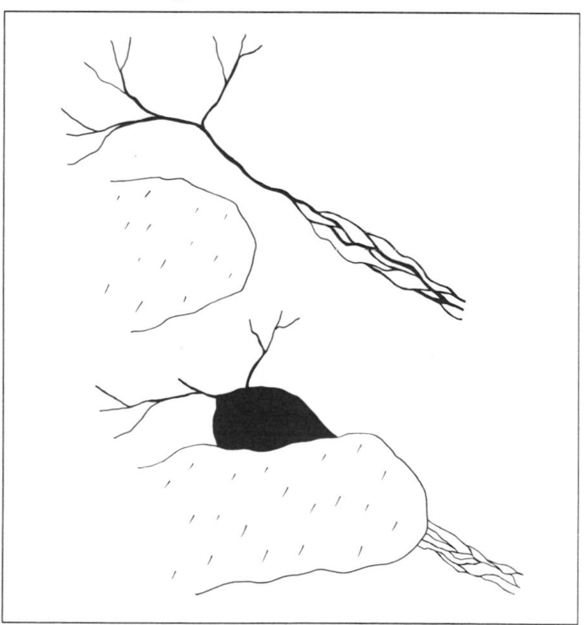

FIG. 7.4—Glaciers that advance across streams can dam them enough to form substantial lakes. The release of these large volumes of water causes outburst flows that are so catastrophic they have a special name, jökulhlaups.

WATER, WATER EVERYWHERE

Water flowing from the mountain glaciers of North America contributes significantly to our overall hydrological budget. In fact, it has been estimated that during any given summer glaciers in Alaska spew over 186,000 billion liters (50,000 billion gal) of water into the stream-flow system. Although glaciers in Colorado only melt 2.5 billion liters (700 million gal) into the summer stream-flow, their contribution is no less important. If you step into the lobby of the Hotel Boulderado, you will see what I mean. They have a handsome old drinking fountain with a sign above stating, "Pure cold water from Boulder owned Arapaho Glacier."

The amount of water held in glacier ice in North America equals or surpasses all the water in all our lakes and streams. If it all melted at once, instead of "trickling" out each summer, the ensuing flood would cause our sea level to rise approximately 20 millimeters (0.8 in)—a small fraction of the melted contributions from Greenland and Antarctica, but no less significant.

CHAPTER

—8—

How Mundane, It's Only a Moraine

I was exhausted. The guidebook had recommended alloting seven days for this trip and we were nearing its completion in the middle of day three. Just one more glacier to cross, a steep, 500 meter (1600 ft) snow field to climb, and we were home.

I stood on top of a ridge of rocks that stretched the length of the glacier. It was a long way down to the ice. Because it was steep and made up of loose rock, dirt, and sand, we decided to glissade down. That's right, a practice usually reserved for pristine fields of snow was now attempted on rock!

I could not believe it, it was *fun*. The additional noise from cascading rocks as we slid down the steep face added to the excitement. It was quick, easy, and my legs had a chance to use new muscles while the tired ones rested.

Once on the ice, we skipped ahead in jubilant assurance that we were near the end of our arduous trip. When we reached the other side we ran into another ridge of rocks, like the one we had just glissaded down. This one, however, had to be climbed up. Ugh.

We spread out, so we would not send rocks down on each other, and scrambled up, often taking three steps just to advance one. After much clattering and jumbling, cursing and swearing, we all made it to the top. Here one of my friends had his camp. He invited us to share some cocoa before we continued our trip home. What a welcome respite. This camp was called Moraine Camp because it was perched on a lateral moraine of the glacier. Now, whenever I think of moraines, I think of that evening sipping cocoa in the cool twilight after an invigorating excursion. The remainder of the climb was made easier.

Pile of Rocks

Although moving ice can sculpt and gouge an amazing variety of geologic features, moraines are unique in their ability to hinder or enhance your progress on and around a glacier. In technical terms moraines are mounds, ridges, or other distinct accumulations of unsorted, unstratified mixtures of clay, silt, sand, gravel, and boulders. As such, they are often called glacial till. In essence, moraines are simply piles of loose rock. They form swirling patterns and stripes on the surface and at the edges of a glacier.

Every glacier you visit should have some type of moraine (see Fig. 8.1). The ice plucks and scrapes pieces away from the surrounding bedrock then pulls it along, pushes it aside, or plows it ahead. Many moraines have their own unique color because the different geologic structures that contribute material to the moving glacier are themselves of distinct colors.

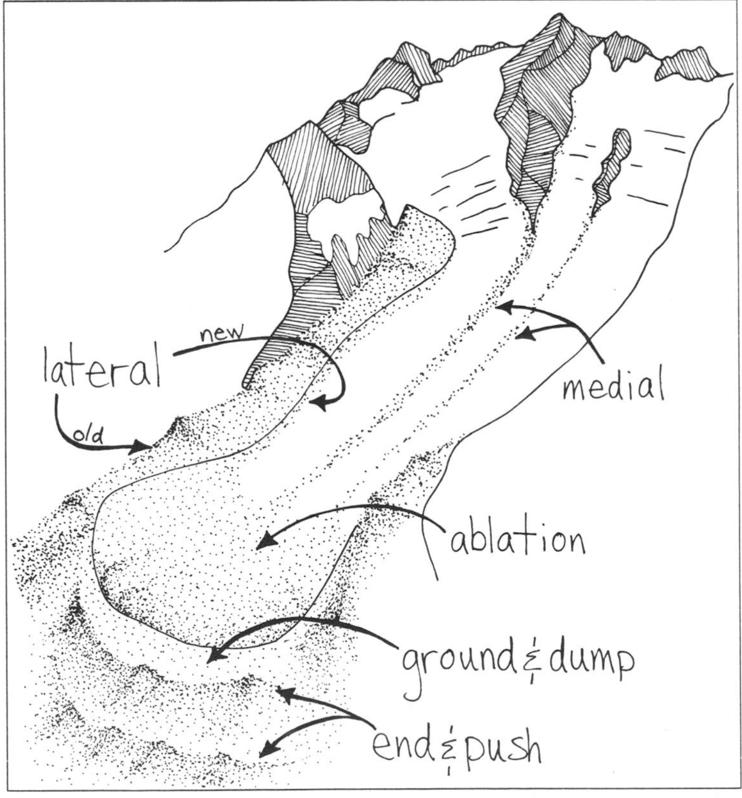

FIG. 8.1—Moraines are piles of rock that have been plucked away from the surrounding bedrock by moving ice. They accumulate on, around, and underneath a glacier.

The lateral moraines that my friends and I glissaded down and scrambled up along the glacier's side edges are typical of nearly all valley glaciers in North America. The rock material for a lateral moraine comes from the valley walls, and as a glacier recedes its lateral moraines become more visible.

The original height of glacier ice was probably at or above the current height of the lateral moraine ridge. This gave me a pretty good idea, looking at ice that was as far below me as the ground would be from the height of a two-story building, that the glacier I was visiting was once quite fat. Naturally, some of the moraine falls away as supporting ice next to the rock wall is removed by the retreating glacier and ice underlying the moraine melts. Therefore, the height of a lateral moraine offers only a minimum estimate of past glacier thickness.

End moraines are similar to lateral moraines in that they are created near the margin of a glacier, this time at the terminus. Probably less steep and less tall than lateral moraines (because ice near the terminus is relatively thin), end moraines are no less interesting. Many times a glacier will have several concentric ribs of end moraines below its terminus. Each one marks a historical position of the terminus.

Sometimes end moraines are also called push moraines. This can be true if an end moraine is displaced by an advancing terminus. However, push moraines are more common in the high arctic. Here, sand and gravel in the out-wash plain ahead of a glacier become cold and brittle. The glacier can subsequently bulldoze through the area, fracturing the frozen sediments into chunky blocks of material that are thrust into arch-shaped ridges ahead of the advancing terminus.

Many times a glacier will simply dump out its supply of entrained rocks as it retreats. The layers or mounds of loose rock from dump moraines can be found all over the ground near the edges of retreating glaciers.

Botanists have become adept at figuring out the age of lichens that grow on the rocks of moraines. In this way they can reconstruct the history of a glacier's advance and retreat. For many glaciers, this allows an estimate of past climate conditions.

For me, medial moraines capture my attention more readily than the other rock piles. These types of moraines create long, dark stripes down the center of the glacier. They usually form at the junction of two ice tributaries where the merging lateral moraines of each tributary form one medial moraine below the junction. If you see a glacier with medial moraines, you may suspect that it is a branched-valley glacier with converging tributaries.

Avalanching rock debris from the steep headwall of a glacier also may create a ridge of till that looks like a medial moraine. However, these are usually less consistent and have a more jumbled appearance than medial moraines that are formed by converging ice streams.

The Saskatchewan Glacier in Alberta has a boldly handsome medial moraine that is sure to beguile you. However, for utterly fantastic stripes, I suggest seeking out some of the larger glaciers that have multiple tributaries flowing into the main stream.

Medial moraines are not only dramatic, they can provide some interesting clues to glacier motion. For example, slow, steadily moving glaciers usually have straight-looking moraines that flow with the glacier parallel to the valley walls. On the other hand, surging glaciers often develop sweeping folds and swirls in their medial moraines much like the swirls and eddies in river water. Surging portions of ice deform moraines with their accelerated motion.

One last variety of moraine begins with those embedded rocks frozen in the ice toward a glacier terminus that cause sparks to fly when skiing over them. These rocks, once a part of the underlying bedrock, are being melted out of the ice at the glacier surface. An accumulation of melted-out rocks will form an ablation moraine. Sometimes ablation moraines are just sparse collections of glacial till. Other times, a thick ablation moraine can completely cover the glacier terminus.

Often the rock and soil debris from moraines, rockfalls, and wind-blown particles pile so deep over a glacier, especially near its terminus, that vegetation begins growing. A few glaciers have full-grown trees thriving in the debris covering its ice.

Although most of the rest of this book concentrates on those parts of glaciers composed of ice, the rock formations of moraines play an intricate role in the complete glacier story. Therefore, I hope you will not curse the next time an ablation moraine takes a chunk out of your new skis, when you have to scramble up a lateral moraine, or when a medial moraine impedes a pristine glacial traverse. Instead, revel in their grandeur and their underlying importance to our understanding of glaciers.

CHAPTER
—9—

Glacier Poison: One Drop and You're Dead

David's mother held my hand while she listened to the story of his death. Climbers who were in the distance had reported seeing David and his rope partner skipping and dancing down the glacier after a successful summit climb. Next time they looked, both were gone. Both had fallen into a crevasse. Both were trapped. Both died. Both are lost in the icy tomb.

She leaned next to my ear. Squeezing my hand so tightly that my fingers began to throb, she whispered, "He told me not to worry. He said that if he ever died in the mountains I should know that he was happy." We both knew he was right, but our tears could not forgive the futility of his death.

That was almost fifteen years ago. As memories of David's accident fade I sometimes regress into my own cavalier methods of glacier travel. But another friend's death or a quick perusal of *Accidents in North American Mountaineering* (published by The American Alpine Club, Inc., and the Alpine Club of Canada) pull me back toward a critical need for caution.

Now that you are familiar with the phenomena of glaciers, how features develop, and where they can be found, the next step is to learn how to avoid their potential dangers. Nothing spoils an exciting glacier journey faster than a harmful accident.

The safety procedures that I have chosen to recommend are only the basics. They require the minimum amount of equipment and, at least for me, are the easiest to remember and perform. There are many other variations on these fundamental techniques. Some require specialized equipment; others require more time to set up or more skill to execute.

Although I believe that there is enough information presented here to help you decide how to stay out of trouble, do not rely only on what you read. It is difficult to describe a complicated technique with only a few words and some pictures. Not only that, but many methods for safe protection work well in some situations and not in others.

The potential dangers of glacier travel are so varied and immense that, for those who truly want to learn how to travel safely on a glacier, I strongly urge you to obtain some practical experience by utilizing guide services and mountain schools. There is excellent instruction that can teach you fundamentals with just a day or two of field work. In addition, many guided mountain trips begin with a day of training.

Once you have learned the basics, it is up to you to practice, practice, practice. None of the protective techniques are easy and the use of ropes, crampons, and ice axes is not always intuitive. Even the most experienced glacier explorers find time to practice safety procedures and check their equipment before each trip to the ice.

Although you will read an occasional *never* and *always* in this chapter, I believe that there are only three *mandatory* rules for safe glacier travel:

PAY ATTENTION,

THINK,

and

BE PREPARED FOR THE WORST.

Other rules, techniques, and suggested equipment depend upon skill level, the kind of glacier, the current season, and one's own objective for travelling on a glacier. The following sections outline many of the expected hazards, when you will most likely run into problems, and how you can learn to avoid trouble.

Slippery Ice

How often have you slipped on ice while casually walking down a city sidewalk during winter? You know that ice is slippery, but did you know that more people die by slipping and sliding than any other glacier hazard?

Although snow and firn are slippery mainly on steep slopes, ice can knock you off your feet even on the flats. The best time to walk on ice is after it has been weathered by a few days of sunshine. This roughened surface may even be navigated with a good pair of running shoes.

The worst time to cross ice is during a rainstorm. The rain washes or melts the weathered ice away, leaving a very smooth, wet, and slippery surface. After a rain, the ice surface may refreeze but remain slick with a smooth, glazed surface.

Often glacier surface conditions change quickly over time and space. For example, a shaded portion of a glacier can remain significantly more slippery than adjacent ice that is exposed to sun weathering. Also, a nice weathered surface in the morning may be converted into a slick ice surface in the afternoon by warm winds or rain. It helps to pay attention to current and changing conditions at all times.

Meltwater streams on the ice can pose additional problems. On large glaciers these streams are formidable rivers. Because the river channel is smooth and slick, crossing can be extremely dangerous. If you lose control while crossing you may never come out.

Be suspicious of mounds of dirt you may see on a glacier. These could be thin layers of moraine covering slippery ice. This condition is especially serious where dirt-covered ice may border deep crevasses. A jump onto seemingly stable rock may offer a surprise slide to danger. Even in tamer conditions, the bloody gash you could get from a fall onto the sharp rocks is bothersome to say the least.

Skis

If you are on skis be sure your edges are sharp. On a steep slope remember to keep your weight over your skis. Do not be drawn into the natural tendency to lean into the hill. This will only remove the force you have on your edges and cause your skis to slip out from under you.

Many skiers use crampons especially made to slip over the ski, under the boot. These types of crampons are called harscheisens and are very useful for steep traverses up hard snow.

Ice Axe

To walk on glaciers, even I—champion of minimal equipment—carry an ice axe at all times (see Fig. 9.1). An ice axe provides a handy anchor for you when crossing steep slopes and can be used to help slow down if you begin to slide.

FIG. 9.1—Parts of an ice axe include the head, shaft, adze, pick, and spike.

Use your axe to help anchor yourself before moving each foot. This will give you a handrail of support in case your new foot position does not hold. When using the spike as an anchor, remember to angle the axe head slightly into the slope (see Fig. 9.2). This should help hold it if you were to lose your balance downhill.

FIG. 9.2—When using an ice axe as an anchor, angle the shaft slightly into the slope to prevent it from slipping out.

In all situations, while walking up a slope or climbing down, be sure you have at least two points anchored and solid before you move the third. For example, make two solid platforms for each foot before you move the anchor of your ice axe, or make a solid platform for one foot and anchor your ice axe before you move your second foot.

On ice, you may want to carve steps with the pick and adze of your axe. Obtain a secure stance with both feet before wielding your axe. As you cut the steps, angle them into the slope to ensure a firm platform. Test each step with your axe anchored before applying your entire weight.

Remember that it is more difficult to climb down than it is to climb up. Although our natural tendency is to face the direction that we walk, many situations call for us to down-climb. Down-climbing requires that you face the slope and walk backwards, downhill. Again you want to ensure that two points are anchored before the third is moved.

At all times, hold on to your axe in a way that will allow you to get into an "ohmygodihopeicanstop" self-arrest position quickly (see Fig. 9.3). One hand is over the axe head with the pick pointed down. The other hand is grasping a comfortable distance down the shaft. The axe is held parallel to the ground about chest high with your elbows bent slightly.

FIG. 9.3—Hold your ice axe about chest high by folding one hand over the head and grasping the shaft with the other hand to be ready to self-arrest.

Self-arrest with an ice axe can be very effective. However, it requires lots and lots of practice. The main thing to remember is to try and get your weight over the ice axe pick as quickly as possible to help drive it into the snow. Speed is important because hazards lurk on a glacier and you want to stop before you reach anything nasty. Always remember that there is a chance the axe could be ripped from your hands. Be certain of your grip before you dig it into the glacier surface, then press it down gradually.

Using the pick of your axe allows you to apply your weight to the entire shaft and confirm a solid hold. If you try using the spike to stop a slide, more often than not it will be flung out of your grasp. Some people find that the adze end of the axe will work best if sliding over loose snow that is on top of an ice surface.

Self-arresting takes practice. You have to learn two things: (1) to stop yourself from sliding into danger, and (2) to control your axe so that its sharp edges and points will not injure you. To practice these "ohmygodihopeicanstop" techniques, find a safe, short, steep slope with a gentle run-out and put on an itsy, bitsy, teeny, weeny, yellow polka-dotted slicker. The slicker will help you stay dry and make you slide faster. Remember to keep your feet up, especially if you are wearing crampons. Now imagine every possible position you may be in during a slip or fall—feet first, head first, on your back, on your front, rolling over, etc. (see Figs. 9.4a, b, c).

FIG. 9.4a—If sliding out of control head first on your belly, try to swing your feet downhill, using your pick as a pivot. Then lift your torso over the axe so you can put your weight on it to help hold.

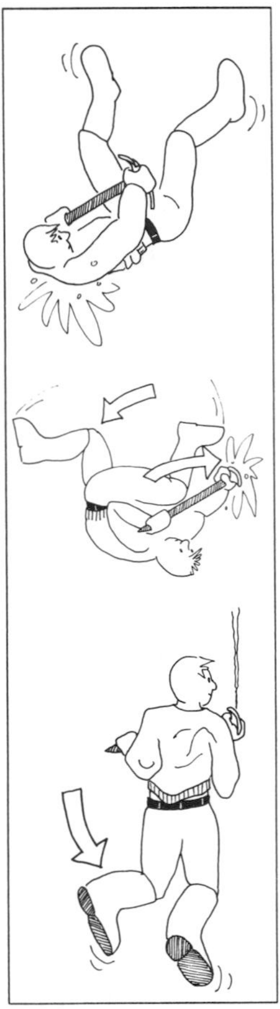

FIG. 9.4b—If you have slipped onto your back, sliding head first, try to reach uphill to plant your pick. Then swing your feet downhill and lift your torso over the axe to slow down.

FIG. 9.4c—If you have slipped onto your back, sliding feet first, rotate your body so you can plant your pick near your shoulder. This should put your torso right over the axe and help you slow down.

Deep, soft snow will either help to retard your slide or begin to avalanche. If you are caught in a shallow avalanche you may be able to dig your axe into a firm bed surface of the slide. Large avalanches pose another problem, which we will discuss later.

Digging an axe into hard ice is very difficult and often ineffective. Many people try, only to have their axe ripped away. It may then flail about causing numerous puncture injuries from its sharp edges and points—the least of your worries as you accelerate down a slick, icy mountain. Remember to get your weight over the pick quickly, then gradually press it into the surface. You may not be able to stop completely but there is a chance that you can slow down enough to regain control.

Crampons

When the ice is too slippery to walk, use crampons. Crampons work best if you can walk flat-footed, with each spike offering a gripping point (see Fig. 9.5). This takes some practice and requires a little bending of the knees and ankles.

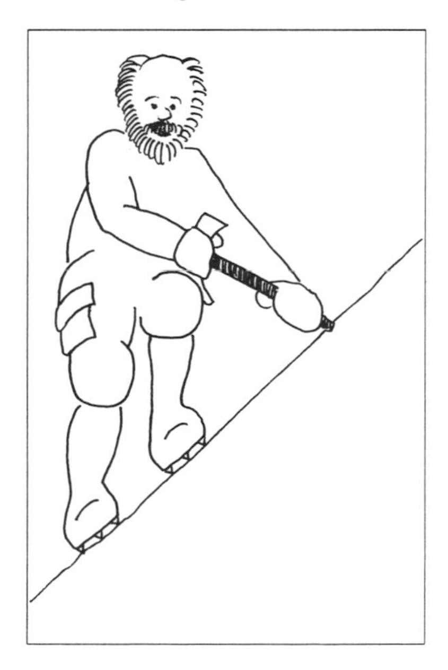

FIG. 9.5—When wearing crampons, try to chose your route so you can walk flat-footed. This will allow you to grip with the most possible crampon points.

The front points of crampons are used for very steep icy slopes and are commonly employed by skilled ice climbers. For this purpose it is important to have stiff boots to spread out the stress over the entire foot. However, it is still safer to use the ten points of a flat foot, and most experienced glacier travellers will plan their route to avoid front-pointing.

For goodness sake, do not wear your crampons on soft snow. They will become crammed with the stuff and behave like a nice slippery ball with your first step back onto ice. If you do go through a short section of snow, check your crampons before stepping back onto the ice. Make sure they remain snow free and effective by tapping your boot with the shaft of your ice axe.

If you fall and begin sliding feet first with crampons on, keep your feet up and rely on your ice axe to slow down (see Fig. 9.6). Many people have broken their legs in fast slides as their crampons obtained the first, sudden grip.

FIG. 9.6—Whenever sliding out of control, remember to keep your cramponed feet up so they will not snag on the ice and flip you over or break your leg.

All too often people gash themselves with the points, trip and fall, or slash a much-needed protective rope by inadvertently stepping on it with crampons. Another common problem is when a crampon breaks or falls off. Usually this happens when the crampon was improperly attached. Sometimes, failure to inspect crampons before using them will cause a defect or loose joint to go unnoticed, causing the crampon to fail as soon as weight is applied.

Before you need crampons, practice walking around with them in a safe place on flat and sloping ice. Before each trip, test the fit and inspect your crampons for any loose or fatigued parts. Sharpen the points, if necessary, with a file.

Heavy Feet

On snow or firn, learn the "glacier walk." Going uphill, this means kicking out steps on your way up. Drive your toe into the slope enough times to build a solid step. Going down or across hills, take big, heavy strides or "plunge steps" that will drive the heel of your boot into the surface (see Fig. 9.7). As you first start out, check each step with your ice axe anchored before you put your entire weight on it. After a while you will obtain a feel for how supporting the surface is. Again, it is important to wear stiff-soled boots for doing the "glacier walk."

FIG. 9.7—Exaggerate your steps while walking on snow and firn so you can plunge into it and ensure a solid platform.

GLISSADE

Some people prefer to descend snow- or firn-covered slopes by glissading, sliding on their feet or seat. To ensure a safe glissade, choose only slopes that you have previously ascended so you know what to expect. Take your crampons off and always, always carry your ice axe in a position that will allow you to use it for holding onto the slope.

Never glissade on ice. It is nearly impossible to stop once you begin sliding on ice. In addition, make sure that the slope you are glissading is free of any ice patches, as each one will throw your balance off enough to lose control. Sometimes a layer of ice is just below the surface. Poke, prod, and test the snow before setting out on a glissade.

Never glissade in an area that does not have enough room to stop before danger. Make sure there are no cliffs or crevasses that will get in the way. It is best to select slopes that have a smooth, concave run-out area, well above the firn limit.

On many slopes, where the surface is relatively soft, you can stand upright while glissading. Hold your axe in a ready-to-arrest position (see Fig. 9.8), and use your boots like skis, keeping your feet together and doing parallel turns to help maintain control.

FIG. 9.8—When glissading on your feet, keep your ice axe in a ready-to-arrest position in case you fall.

In other cases, like on very steep slopes or with snow that is too firm to hold an edge, you may want to utilize a crouching glissade. By spreading your feet apart into a comfortable stance and leaning back on the shaft of your ice axe, you can career downhill with the greatest of ease. Have one hand over the head of the axe and the other near the bottom of the shaft (see Fig. 9.9).

FIG. 9.9—For crouching glissades, grip the shaft of your axe close to the spike so you can apply a strong lever to help control your speed.

Those who do not mind a wet bottom, or who have on waterproof pants, may be more comfortable performing a sitting glissade. Here, again, you want to hold one hand over the head of your axe and the other near the bottom of the shaft (see Fig. 9.10). Use the spike to control your speed. If you get going too fast, roll over with your chest over the shaft and use the pick to help stop.

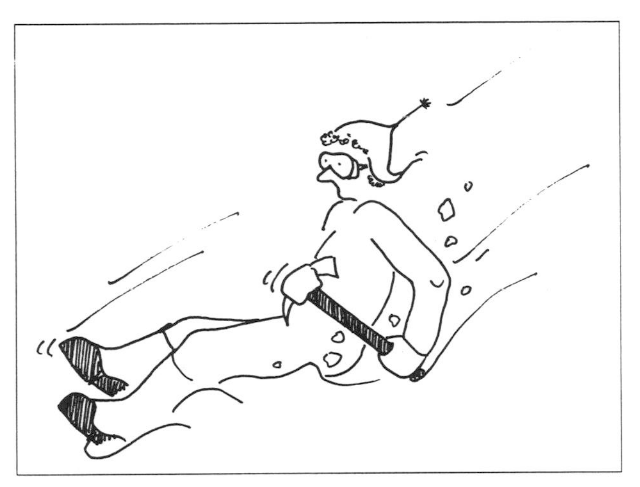

FIG. 9.10—Grip the shaft of your axe close to the spike for sitting glissades also. It is more difficult to leverage in this position so if you find yourself in trouble, roll over to arrest with the pick of your axe.

Rope Together

Another protective measure to inhibit an out-of-control slide is to rope up. There are several methods of tying yourself into a protective length of rope. Some books recommend tying the safety rope around your waist. However, if you have ever dropped suddenly into a deep hole with a rope tied around your waist, you will know the pain of being cinched around your diaphragm. People can suffocate quickly in this position.

At the very least a seat harness should be used to attach yourself to a protective rope. Although this does not prevent the potential for ending up upside down, it should help avoid pressure on your diaphragm. For those who are serious about a long and healthy life, the combination of chest and seat harness is recommended.

I have seen as many as fourteen people tied onto one rope. Can you imagine what would happen if thirteen people started to slide out of control and you were the only one left to stop them? On top of that, the more people tied to a rope, the more difficult it is to keep taut. Rope teams of two or three are very effective, and four is the working maximum.

If on foot or snowshoes, a popular distance between rope partners is a taut rope of about 10 meters (30 ft). The extra rope can be coiled diagonally across each chest of the two end partners and incorporated into their harness systems (see Fig. 9.11). If skiing, you need lots of room to turn. A useful length between ski partners is about 25 meters (80 ft) to allow some room for turning while keeping the rope tight between skiers (see Figs. 9.13a, b).

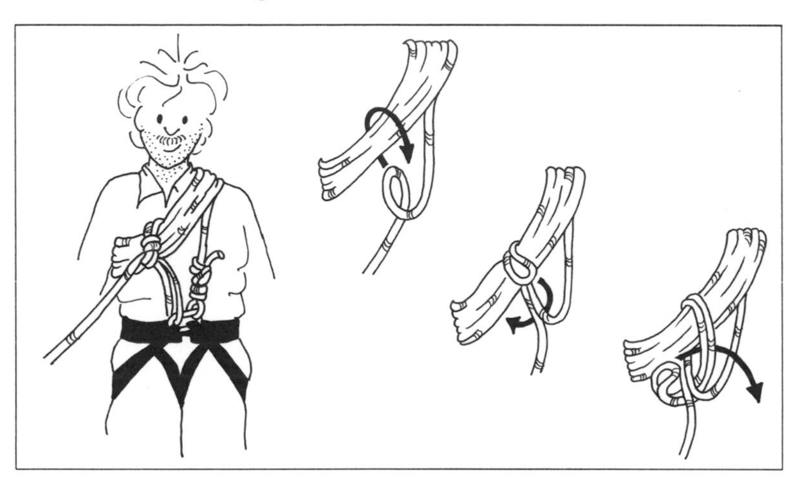

FIG. 9.11—Two walkers on a rope may want to shorten it. You can use the extra rope as a chest harness by wrapping a coil over one shoulder. Tie the coil with a simple overhand knot then attach the loop to the carabiner on your seat harness.

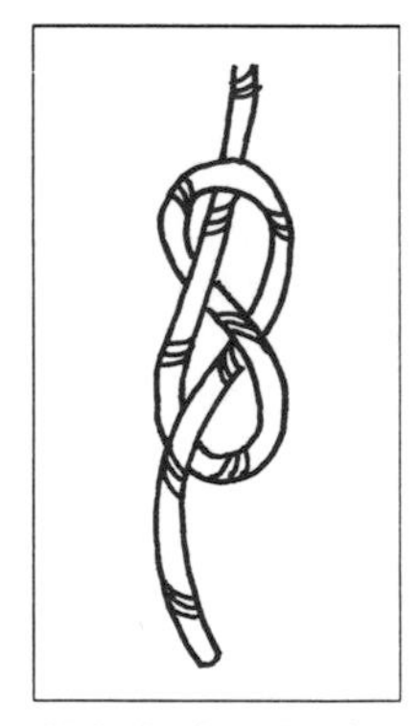

FIG. 9.12a—To tie a figure-eight knot, form a loop in the rope then bring the end back behind, over the top, then through the loop to form the shape of a numeral eight.

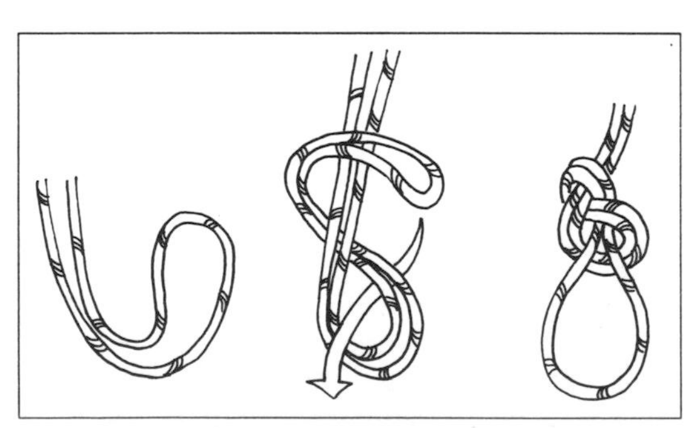

FIG. 9.12b—A figure-eight loop is like a figure-eight knot except that it is made at the fold of a rope to form a loop.

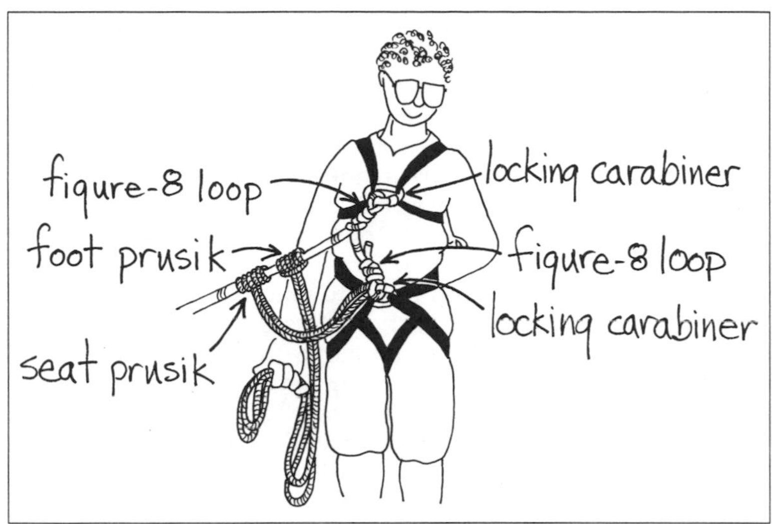

FIG. 9.13a—Two skiers, or two walkers with a short rope, will want to use the full length. The end of the rope should be secured to the carabiner on your seat harness with a figure-eight loop. Another figure-eight loop should be tied in the next short section to attach to the carabiner on your chest harness. Your foot prusik should attach closest to you with your seat prusik away from you.

FIG. 9.13b—If you are in the middle of the rope, you may want your seat prusik tied onto one side of the rope and your foot prusik onto the other. This will help you react to a variety of situations, whether you, the leader, or the trailing partner falls first.

In all cases, be sure to keep the rope taut. Once a slide begins, the eventual acceleration can develop tremendous force. A slack rope will only enhance this force. Never carry extra coils of rope in your hand.

If you see or feel your rope partner falling, your first reaction is the most important. Your strongest efforts should be made to prevent yourself from being pulled over backwards.

A figure-eight knot in the rope (see Fig. 9.12a) or a secured prusik (see later section) held in your free hand will help you yank on the rope if you see your lead partner begin to slip. More than once this has been able to stop a potential slide.

To help hold on in the event of trouble, if you are on snow, it should be possible to drive the shaft of your ice axe up to the hilt as you plunge onto the handle. Try to tilt the shaft slightly as you drive it into the snow so that the handle is leaning toward the slope, away from your falling partner. This will make it more difficult for the axe to be pulled loose.

If your shaft cannot penetrate a significant length, it will rip loose with the force of your falling partner. Therefore, be sure to test the penetrating ability of your ice axe frequently. However, if you do find yourself on hard firn or ice you may have to use the "ohmygod-ihopeicanstop" method to lock the pick of your axe into the surface with your torso on top as best you can.

On skis your best maneuver will be to shift your edges perpendicular to the fall and drive them into the surface (see Fig. 9.14a). You may be able to hold the fall by staying upright with all your weight on the edges of your skis. If you end up dropping and holding with your entire body, use your skis as plows and, with your weight near the bottom of your poles, press their sharp tips into the slope (see Fig. 9.14b).

FIG. 9.14a—A skier on steep slopes or ice should be ready to arrest by gripping both ski poles with the downhill hand near the handles and the uphill hand near the tips.

FIG. 9.14b—If you begin to slide while on skis, move your body over the tips of the poles to help them grip the slope. Use your skis like plows to help stop.

Fixed Belay

Keep in mind that more than one person can lose their footing at the same time. If there is a good possibility of slipping, especially if negotiating steep and/or dangerous terrain, or if you are much smaller than your partner, then it is best to travel with fixed belays.

During a fixed belay, one partner will advance while the other partner controls the rope and prepares to brake in case of a fall. The object is to maintain a taut rope so that any slip by the advancing partner can be controlled immediately. If there are more than two people on a rope, only one should be moving at any time. The remaining partners should each hold a secure belay. In addition, it is important that the one(s) controlling the belay be in a safe position with a secure anchor.

There are many different belay techniques. Each has its own advantages and disadvantages. The safest techniques are usually the most time-consuming. With special equipment you can improve any belay substantially.

In this section I will describe one quick belay that may assist you in negotiating snow slopes. Later on we will learn about some more substantial belays that help protect against vertical falls or falls on ice. Once again I implore you to experience different belaying methods by taking a practical lesson from a professional, then practice and practice and practice some more.

Boot-Axe Belay

In snow, one method of belaying is to wrap the rope around the shaft of your axe and your boot into what is called a boot-axe belay. The belayer should first stomp a firm platform in the snow, facing sideways along the slope with the uphill foot slightly ahead of the downhill foot. The stomping part is important. You need to compact the snow as much as possible to provide a solid anchor.

After you have created a solid platform for both feet, drive the shaft of your ice axe vertically into the snow on your uphill side. Keep the axe head perpendicular to the slope so that the wide cross-section of the axe can be used to improve its hold. Hold your uphill foot against the axe head.

The active side of the rope (that is the one your friend is tied to) should go over the top of your boot to prevent it from digging into the snow. In an S pattern, loop the rope under the axe head, around the uphill side of the shaft, back over your boot, then around your ankle.

The active rope should be kept taut. The inactive rope (the side that is tied to you) should be in a loose coil below your downhill boot. Be certain that it is not between your legs or wrapped around something that could jam the system or pull you off your feet. Also be sure that it will not tangle as it slides through the belay anchor.

Your uphill hand has two functions. One is to guide the active side of the rope and the other is to hold the axe secure in case of a fall. The guiding hand should ensure that the rope is kept underneath the axe head. In the event of a fall, that hand should move quickly onto the axe head to help hold the axe in place. Do not forget this. The force of a fall can rip the axe out of the snow or break it. You must hold it with all your might.

Your downhill hand has only one function, that is to act as a brake. Prevent your breaking hand from getting pinched against your boot by holding the rope with enough room to wrap the rope against the back of your boot. Never let your braking hand leave the rope.

If your partner is travelling downhill from you, braking should be straightforward (see Fig. 9.15a). Your uphill hand should quickly move to secure the axe head. Your braking hand should grab the rope and bend it around the back of your ankle to increase friction on the rope and help stop it.

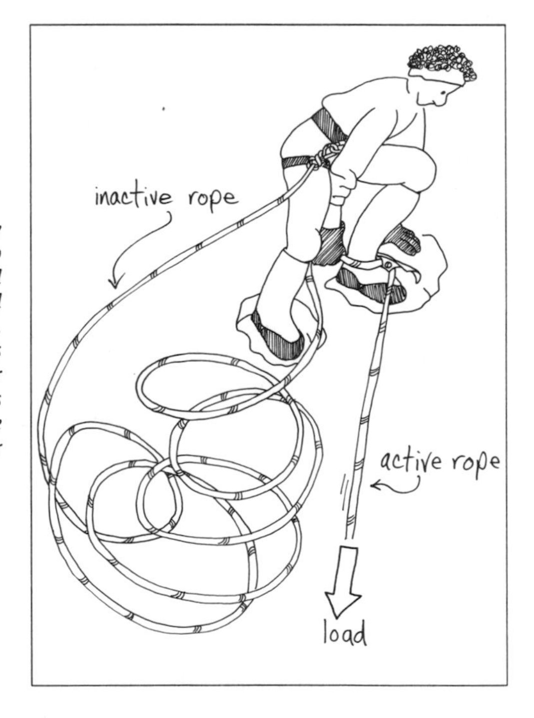

FIG. 9.15a—In firm snow you can use your ice axe to help belay. Your uphill hand is used to guide the rope and hold the axe in case of a fall. Your downhill hand applies the brake and should never leave the rope. Wear gloves and make sure that the rope is not tangled around your legs.

If your partner is travelling uphill from you, braking is less obvious (see Fig. 9.15b). Because there will be quite a bit of slack created as your partner slides past you, the force of his or her fall could become tremendous. If you try to brake this fall suddenly, the increased force could rip out the whole system, you with it. For this reason, braking should be gradual. With your uphill hand, be sure that the rope does not flip out from underneath the axe head. Let the rope slide through your brake hand as you guide it around your ankle. This should allow the rope to reach a gradual stop.

To always have enough room for a gradual stop it is important not to belay out the entire length of rope. Move to the next belay before the last few meters of rope are played out.

Move diagonally while travelling on belay upslope. This will prevent the leader from falling behind the belayer. Also, keep your belay pitches short, less than 20 meters (65 ft). The boot-axe belay is not strong enough to support hard falls, and short pitches can prevent this from happening.

FIG. 9.15b—If using a boot-axe belay while travelling uphill, remember to keep the pitches short and traverse at an angle to the fall-line so you don't end up behind the belayer if falling. To ensure plenty of rope for a gradual brake, the belayer should not play out the entire length.

Do not forget to wear gloves on both hands. And remember to keep the rope against your boot as you wrap it around your ankle. Spencer Tracy may be able to hold onto a rope with his bare hands to stop Robert Wagner from falling off a mountain or to catch a fighting marlin, but the average person cannot. As a belayer, the very end of the rope is tied to your harness. If the braking system fails then it is you who will be pulled into trouble next. Please remember to practice under realistic circumstances. Ask your biggest friend to help simulate falls. Find a safe place to practice so that when things do fail no one will be injured.

Slots, Slots, Slots

Crevasses pose the second most commonly fatal glacier hazard. Usually wedge-shaped, victims can easily fall into the wide opening at the surface then jam between the walls that pinch closed near the bottom. People have suffocated as their lungs are crushed between the walls. Others die of exposure (hypothermia) before attempts at extricating them can be successful. Some rescuers in Europe have been known to use chain saws to cut wedged victims out of a crevasse.

Avoidance

The best way to avoid hassles is to not fall into a crevasse, which is a trick in itself. One way is to avoid areas with numerous, closely spaced crevasses. Learn where the most likely places for crevasse formation exist (see Chapter Six), try to obtain a broad look at the glacier's crevasse patterns, and simply stay away from the heavily cracked areas.

Often the best route onto a glacier is where it is relatively flat. At a glacier's edge you may be able to see the marginal crevasse pattern. If so, you might find an unobstructed route by walking a line about 45 degrees from the edge and moving parallel to the crevasse pattern.

Once on a glacier the best route up or down is usually smack dab in the middle, where there are the fewest marginal crevasses. However, be aware of transverse crevasses, especially in the accumulation zone and wherever the slope steepens.

Even if you think you have learned a glacier's crevasse patterns pretty well, remember that glaciers move and change and may not have the same patterns year after year. Therefore, restudy glaciers with each return visit.

Obvious deep gashes in the ice surface are easy to see and avoid. Crevasses are most visible during summer in areas below the firn limit where there is bare ice and no surface snow to hide the slots.

Above the firn limit, in the névé portion of the glacier, the best time to travel is during the summer after the surface snow has broken apart or melted enough to reveal the crevasses. Nearly all crevasses are open and visible by July in the Lower 48, and later in the summer season as you move northward into Canada and Alaska. But do not count on this as a precise clock if there has been an unusually cool summer, or if you are in an area with a particularly low snow line. You should avoid travelling on névé during fall and early winter when the snow blanket covering slots is particularly fragile.

Anywhere there is snow or firn on a glacier, especially during seasons with new snowfall, crevasses may be hidden by a breakable white blanket. Sometimes a subtle dip in the snow surface or a thin crack will indicate the presence of a hidden slot. Often there is no indication at all. Sir Edmund Hillary, in his book *No Latitude for Error,* said that many times on his way to the South Pole their first indication of entering a crevasse field was when their Ferguson (*ahem*) tractors broke through a snow bridge.

During mid-winter glaciers are good skiing terrain. A deep winter snowpack can build up over crevasses and make relatively strong bridges. However, you should try to stay in the middle and on the gentler portions of the glacier where the crevasses will not be large. If you are skiing perpendicular to the crevasse pattern, your skis can help bridge some of the unsuspected narrow crevasses in case the snow cover does break away. The steeper portions of a glacier threaten not only with larger crevasses but also with snow avalanches and are best to avoid.

If you are suspicious of a crevasse's whereabouts poke your ice axe into the glacier surface before each step (see Fig. 9.16). If there is easy resistance or if it breaks through you may have found a false edge; that is, one bridged over with weak snow. Find another place to walk. Remember that some coverings are not uniform and just because your ice axe is resistant to penetration does not mean that everything is completely safe. Stay on guard at all times.

FIG. 9.16—Use your ice axe to feel for a change in resistance in the glacier surface and possibly detect hidden crevasses.

If you can see that you are near a crevasse, walk around if possible. When the crevasse is too long to circumnavigate it sometimes is easier to find a narrow spot and just jump over the thing (see Fig. 9.17). Make sure the slot is completely visible so you know just how far to jump. Check each edge so you do not take off or land on an unstable overhang. If you are like me, you will want to remember not to look down as you jump because it can be a frighteningly deep hole that you are flying over. Keep your feet up so you do not trip and tumble when landing on the other side.

If you are not familiar with your jumping ability, practice first, especially if you are wearing a pack. If you have on a heavy pack, take it off first then belay it over after you have safely reached the other side.

FIG. 9.17—Walk around crevasses if possible or jump over a narrow portion.

Also, hold your ice axe in the self-arrest position. If you do slip, you can thrust the weight of your body (your torso) over the pick. If you are lucky and have practiced this stuff before you may be able to stop yourself from sliding past the edge.

Moats can be dangerous slots as well. To avoid falling into one of these use caution as you approach the rock edge of any glacier. If already on the glacier, poke your axe into the snow in front of you to test the strength of snow that may be overhanging a moat. Be sure to provide plenty of clearance for a take-off and landing if you need to jump over a moat.

Rope Together

The best protective measure when travelling through a crevasse field or when travelling on the névé portion of a glacier is to use a protective rope and keep it taut. This way, if you do fall into a crevasse you should not drop very far. A tight rope will minimize the amount of momentum that can be gained during a fall and help you stay alive.

Many people suggest walking perpendicular to the crevasse pattern. Others suggest walking parallel. Often you cannot see the crevasses well enough to determine if you are indeed perpendicular or parallel. However you position yourselves to travel on a glacier, just remember to stagger yourselves enough so you do not all fall into the same slot at the same time.

If you are following your rope partner, use a secured prusik (see later section) or tie a figure-eight knot in the rope to hold onto with your free hand. In this way you may be able to jerk the rope back and stop the leader if he or she begins to fall. The ice axe in your independent hand can be used to start an anchor.

Leaders who feel a rope partner falling behind them must establish an anchor immediately. This means jamming the full length of your axe into the snow with your torso, or falling into a self-arrest position with the pick dug into the firm snow or ice.

Crossing a Snow Bridge

If you have found a snow bridge to cross, inspect it thoroughly by looking at it from every possible angle. What looks thick and strong on one side may be hollow on the other side. Use your ice axe as a probe pole in front of you to make sure that the bridge is firm and you are starting out on solid ground.

If the strength of a snow bridge is suspicious, crawl across it. This will help to distribute your weight and not unduly stress an individual portion of the bridge. Remember that a bridge crossed in the morning may have softened and weakened by afternoon warming. Do not expect to be able to follow the same route on a glacier from dawn to dusk or day to day.

Use a fixed belay when crossing snow bridges. Although risky, sometimes a boot-axe belay will work. However, when belaying across a large snow bridge, the traditional waist belay is most effective.

Waist Belay

In a waist belay, the belayer wraps the rope around his or her waist, uses one hand to guide the active side of the rope and the other as a brake. The braking hand should never leave the rope. Hold it out, in front of your body, with a slightly cocked elbow. This will allow you to draw it across your belly and brake against a fall without it becoming pinched to your side or pulled painfully behind you. Some suggest taking an extra loop around the forearm of your braking hand to increase the amount of friction available for braking. However, wrapping an arm this way will increase the likelihood that it will be pulled and trapped behind your back in a full-force fall.

The belayer should establish firm ground by stomping out a place to sit with solid footholds. Sit on a foam pad or something to protect you from the cold, wet snow. Keep plenty of room behind you so that you can maneuver the arm that holds the active end of the rope (see Fig. 9.18).

Anchors

A strong anchor should be set behind the belayer. To prevent undue motion away from the anchor if the load becomes excessive, place it as close behind you as is comfortable. The belayer is then tied to the anchor, either using a length of rope from the inactive end or tying on with a spare section of 9 to 11 millimeter rope or 25 millimeter (1 in) webbing.

Although it may be more effective to tie extra webbing from the anchor to the back of your harness, it is possible to clip in to the anchor

FIG. 9.18—When performing a waist belay, make sure that you are securely positioned with good footholds, a back-up anchor, and enough room to maneuver.

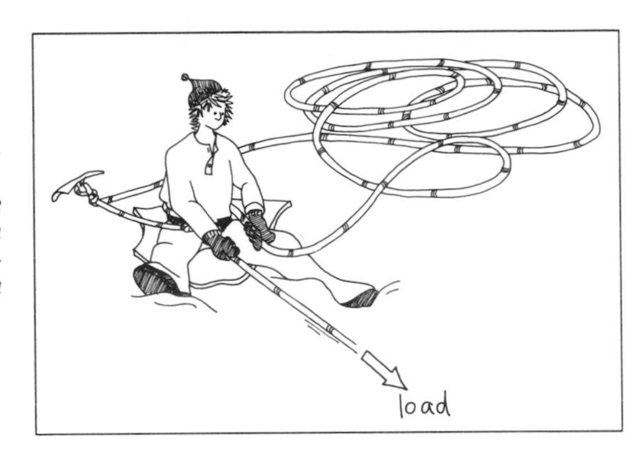

from the inactive side of the rope that is already attached to the *front* of your harness. This method is quicker, but it will cause you to rotate more strongly with the force of the belay. Therefore, bring it around behind you on the same side of your body that holds the active side of the rope. Tie a figure-eight loop in the rope where it will attach to the anchor with a locking carabiner.

In firm snow or firn an ice axe, strong shovel, or skis may provide a reasonable anchor. An ice axe should be pushed nearly vertically its full length into the snow and a slight angle away from the belay load (see Fig. 9.19a). A shovel should be pushed vertically into the snow its full length (including scoop and handle) with its curvature concave away from the belay load (see Fig. 9.19b). Skis must be inserted as far as possible, slightly angled away from the load. A mat should be placed around them to protect the anchor rope from being cut by sharp ski edges (see Fig. 9.19c).

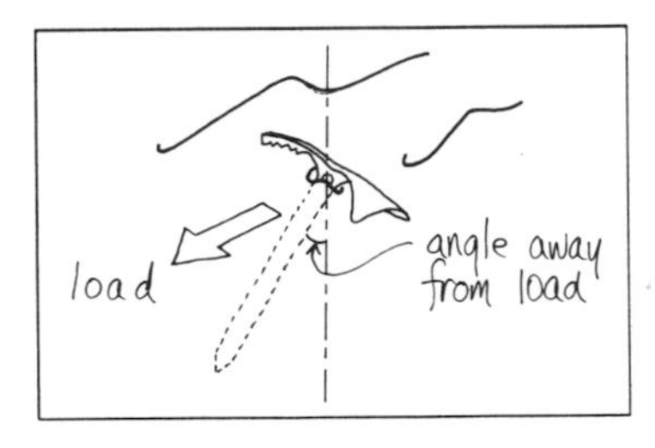

FIG. 9.19a—It is possible to use an ice axe as anchor if the snow or firn is firm and you can drive the shaft in up to its head. If there is a layer of loose snow on top, scrape it away to reach firmer layers underneath. Angle the shaft away from the load to reduce its chance of slipping out.

FIG. 9.19b—A shovel also can act as anchor if the snow or firn is firm and you can drive the handle in to its full length. Again, if there is a layer of loose snow on top, scrape it away to reach firmer layers underneath. Insert the shovel so that its back faces the load. This will allow its curvature to help reduce the chance of pulling out.

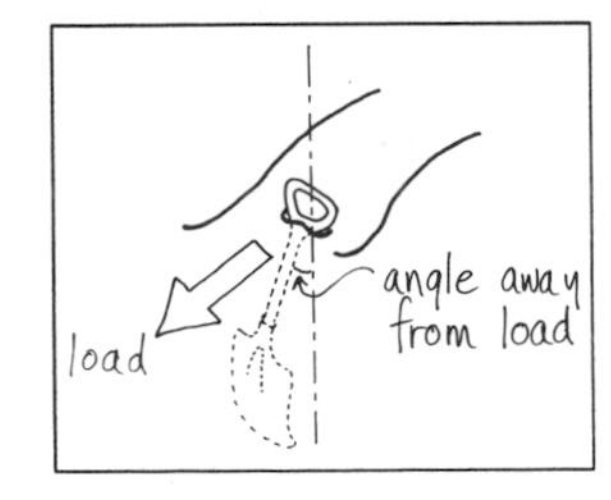

FIG. 9.19c—Skis too can be used as anchors in snow or firn. Drive them as far as possible. Loop the rope next to the glacier surface, putting a pad around the ski's sharp edges to protect the rope.

In deep, soft snow, a dead-man anchor should be used (see Fig. 9.20). Dig a deep hole. Tie everybody's pack together to make as heavy a load as possible. Attach the whole bunch to your anchor rope then cover it up with snow. Stomp all over it and cover it some more until you are sure that it is under a strong load of snow.

FIG. 9.20—In soft snow, use a dead-man anchor. Dig as deep a hole as you can, then pile as many heavy things into it as possible. Loop the rope around the whole mess then fill in the hole and stomp it down.

In ice you can shape a bollard to help anchor a belay (see Fig. 9.21). The size of a bollard depends upon the strength of the ice. In strong ice the bollard should be at least 40 centimeters (16 in) in diameter and at least 20 centimeters (8 in) deep. Make the bollard larger and deeper if the ice seems to be relatively weak.

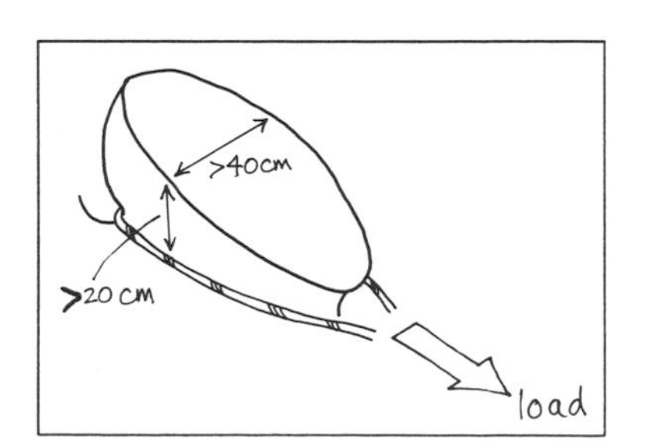

FIG. 9.21—In ice a bollard works well as an anchor. Make it at least 40 centimeters (16 in) wide and 20 centimeters (8 in) deep. In weak ice or firn, make the bollard bigger. Notch the back to prevent the rope from slipping.

Obviously any available rocks may also act as good anchors. However, remember that not all rocks are as solid as they first appear, especially on terrain surrounding an active glacier. I have pulled out Volkswagen size boulders with a flick of a wrist. Inspect each rock before relying on it as an anchor.

Prusik Slings

Even the best protection against falling into a slot may not always work. It helps to have the ability to climb out of trouble. For this purpose, prusik slings are the mountaineer's answer to an elevator. They can also be used to help secure an anchor in case you are the one left on the surface.

The most popular type of elevator these days is to have one prusik for your seat and one for both feet. Tie the seat prusik onto the rope farthest away from you, clipping the other end to the carabiner on your seat harness. Attach the foot prusik onto the rope closest to you. At the other end of the foot prusik, tie a slip knot that can be used later. Use a three-loop prusik knot to tie each sling onto the rope (see Fig. 9.22).

If you are at the end of the rope (literally, not figuratively) and have some extra coiled around your chest, make sure that the prusik knots are attached to the part of the rope that extends to your partner, not the coiled end. Although it may seem obvious to you as you sit in your living room, attaching to the correct end of the rope is not always straightforward while in the fits of a snowstorm. Just remember to double check and you should be fine.

If leading, stuff the end of the foot prusik loosely into a pocket to keep it out of the way but handy if needed. If trailing, hold the end of the foot prusik in your hand to help pull on the rope in case your partner begins to fall.

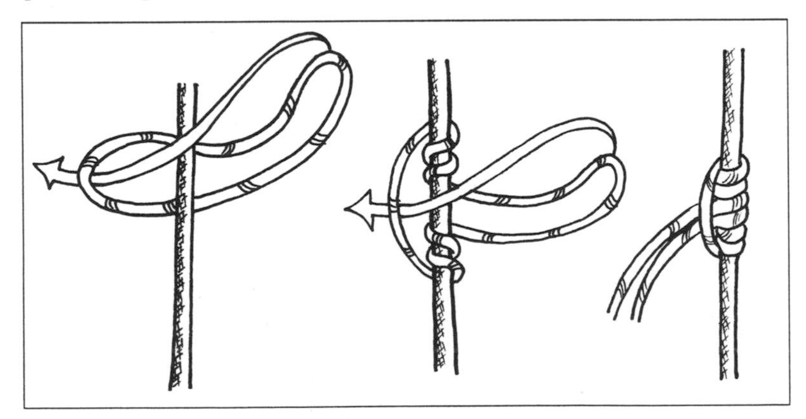

FIG. 9.22—Prusiks are like slip knots. When pressure is applied they grip, when slackened they slide. Three-loop knots wrap the prusik three times over the climbing rope.

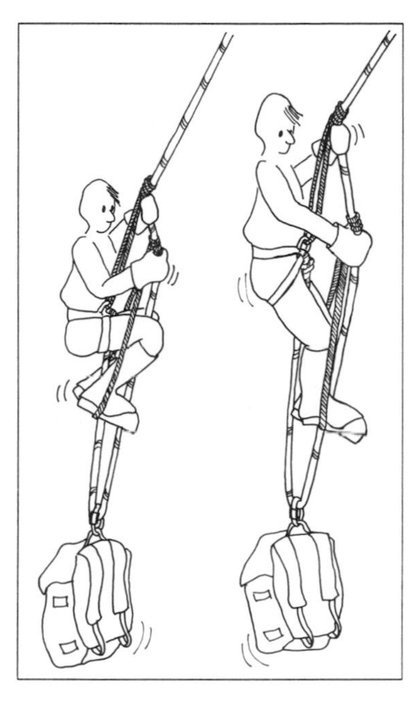

FIG. 9.23—To prusik out of trouble, take your pack off and clip it to the climbing rope between your seat harness and prusik knots. Slip your feet into the foot prusik loop then bring them up underneath you while keeping your body close to the rope. Loosen the foot prusik knot and slide it up the rope. Now stand and push the seat prusik knot as far as possible up the rope. Repeat until outta there!

If you are in the middle of the rope, which side to attach your prusiks is not obvious. It may be a good idea to attach one prusik to the rope in front of you and the other to the rope behind you. Although you will always have to move one prusik before you can start to climb, at least there always will be one available for you to use no matter who falls first.

Climbing Out of Trouble

When you first fall into a crevasse you may be a little disoriented; it could be dark and most certainly wet. Hopefully your rope partner was alert and prevented you from falling too far. If possible, look for a ledge or something you can rest your weight on while those friends who remain on the surface can secure an anchor and belay.

When the coast is clear for you to begin squirming about, remove your pack. With a carabiner, clip your pack onto the rope between your harness and the first prusik knot. This way it will dangle below you when you climb and stay out of your way. The dangling pack also helps to stretch the rope, making it easier to move your prusik knots and climb.

Next, put your feet into the prusik loop. Cinch the loop over your boots to prevent your feet from slipping out. Stand up with your weight on the foot prusik. You will have more control if you keep your weight close to the rope by bringing your feet up underneath you as you prepare to stand. Loosen the seat prusik knot and move it as high on the rope as possible. Now shift your weight to the seat prusik and move the foot prusik. Repeat these steps until you reach the lip of the crevasse (see Fig. 9.23).

At the crevasse lip, you may need some help from your friends on the surface to overcome the overhanging snow that often exists. Whatever you do, do not panic. Try to move methodically to save energy. If you can reach your shovel to dig away the loose snow, you may find the going a bit easier. But for goodness sake, do not hit the rope with your shovel blade.

It helps to practice prusiking. You can do this by tying an old rope to a tree or a rafter in your garage. Use the same kind of rope that you would carry with you on the glacier. Attach your prusiks to the rope and climb up. By practicing you will be able to find the optimal length of prusik loops to use as well as fine-tune your technique for fast climbing. Try it also with a full pack, just to add some realism.

Holding On in the Event of Trouble

If your friend has fallen into trouble, your goals as a holder are (1) stop yourself from being dragged into the hazard with your friend; (2) create a fixed belay that will hold itself without you; and (3) provide your fallen friend with the necessary tools to extricate himself or herself, or rescue your friend if injured or unconscious.

As the holder with your ice axe shaft already driven into the snow, you should be able to secure a belay. If the snow is stiff, and the axe feels secure, wrap the loop of your foot prusik around the axe handle just under the head.

If your ice axe is plunged into soft snow, be aware that it may pull free if you remove your weight from it. It may be possible to wrap a prusik loop around the handle, then sit or stand on the axe head while you build a dead-man anchor with your pack.

As the holder on skis in snow, you can drive the shafts of each ski pole into the snow. Remember to tilt them slightly so the protruding ends are away from the taut rope. This will increase their holding power. The old-fashioned smooth-handled grips can be driven pretty easily. Many new ski poles have grips and baskets that can be removed before pushing the shafts into the snow.

Wrap the loops of your foot prusik around the protruding ski pole ends. Test the holding ability of these poles. Once you are convinced they will hold, you can remove your skis and set them as more secure belays.

Hopefully there will be other people around to help secure a belay. Remember to test the hold of your anchors before you shift your weight.

Preparing the Belay

If the fall has taken place in snow, the rope will cut through the crevasse lip and make it very difficult to climb out. A spare rope, or spare end of a long rope, can be used to alleviate this problem. Your partner can use this spare rope to climb his or her way to safety, or you can use this rope to lower yourself and rescue an incapacitated partner.

Anchor one end of the spare rope. Place a pack, ice axe, ski, or ski poles on the lip of the crevasse and tie a safety rope on to them so they do not become crevasse victims themselves. Throw the loose end of the spare rope over this lip protection. Do not approach the crevasse lip unless you are on belay yourself, either from another friend or with a rope attached to an anchor.

Once a belay is set you should communicate with your fallen partner. Hopefully, he or she will be able to finish the job of extricating himself or herself from this point.

Rescuing a Friend

If your crevasse-bound partner is injured or unconscious, the problem becomes more serious. Remember, you are your friend's best hope for survival. By the time a rescue party can reach the site your friend may freeze to death or die of injuries. Do not delay.

As above, your first priority is to secure yourself and fix a belay to hold your fallen partner. Then use your extra rope to belay yourself or another free party member to the crevasse edge to assess the situation.

If your unconscious friend is dangling freely with a seat and chest harness, it should be possible to haul him or her to the surface. In addition to placing some lip protection to prevent the rope from cutting into the edge of the crevasse, you will need to rig a pulley system to help you haul.

Like all other rope techniques, there are several different types of pulley systems available and some pulley hardware that can help quite a bit. The following describes a basic Z-pulley. Although this system is helpful, it is still very difficult to haul anybody out of a crevasse. You may want to learn other, more complicated techniques that could improve the hauling ability of this system by taking a practical course in crevasse rescue.

To build a Z-pulley, first secure a second anchor behind your initial one (see Fig. 9.24). (Remember to use ice axe, skis, or shovel anchors only in hard snow or firn. Use dead-man anchors in deep, soft snow and bollards in ice.) The prusik that you used to secure the rope onto the first anchor will now be used to prevent the rope from sliding toward the crevasse as you haul. On the anchor side of this prusik knot, attach a carabiner that then attaches to the first anchor.

Loop the rope back toward the crevasse and attach it to the rope with a second prusik sling. Shorten this prusik as much as possible. To complete the system, you should be attached to the first anchor with enough room to reach the crevasse lip.

As you haul, each prusik will pull toward the anchors. You may need to periodically move them back toward the crevasse. Slow down as your friend reaches the crevasse lip. It could be very difficult to get him or her over the lip, so take your time and work methodically. Practice hauling techniques with a competent instructor in a safe place.

Saving a friend's life depends upon your being creative, not panicking, moving methodically, and testing each belay before relying on it.

Travelling Alone

I cannot advocate travelling alone on a glacier, but people do. A method of self-protection that I have heard about is to hold a long pole under your arm. If the pole is perpendicular to the prevailing crevasse pattern, it may bridge the gap if you happen to fall in. I do not doubt that this is possible since my skis have offered a quick bridge many a time when small slots have opened suddenly beneath me. Be that as it may, let me remind you of a popular poem.

Sir Archibald Bone
Climbed glaciers alone.
"I find people boring," quothe he,
"It's annoying to walk
And hear people talk."
He was fond of his own company.

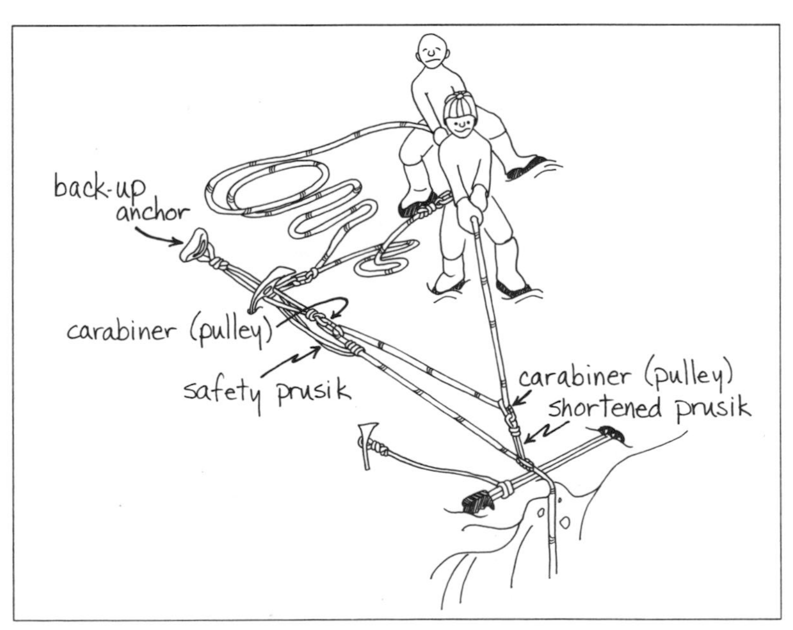

FIG. 9.24—The Z-pulley works best if you have several people to help pull. Add a back-up anchor for the whole system and to anchor your lip protection. Each rescuer should be anchored as well.

So off he would go
Up ice, over snow.
Like a fly on a huge bowl of porridge.
'Til one day, alas,
He slipped down a crevasse,
And he's still somewhere deep in cold storage.

Snow Avalanches

Snow avalanches can happen at any time of year and any time of day or night. Too many mountaineering books claim that spring snow avalanches only happen on sunny afternoons and winter snow avalanches only happen during storms. This is wrong.

It is much too easy to mistake old, unstable snow for the more stable conditions of firn. Also, it is ineffective to guess snow stability by just looking at the surface conditions. Any snow-covered slope has the ability to slide. You need special skills to learn where and when snow avalanches may occur, and these are not easily taught. Many books will provide "rules of thumb" to explain avalanche behavior. I will not. I have witnessed too many fatalities where people got into trouble by following a "rule of thumb."

The easiest way to avoid snow avalanches is to stay off steep snow-covered slopes and out of gullies and to not stop underneath them. Avalanches can travel long distances over bare ground. Therefore, pay attention to what is above you even if you are a long way from the snow and glaciers at higher elevations.

Unfortunately the smooth, snow-covered portions of glaciers and surrounding terrain are often the best places to travel. Certainly they make the best skiing slopes. Always to be prepared for snow avalanches—they can occur at any time of the year. Here are a few tips.

First, learn to recognize conditions that cause avalanching. Because so much of avalanche recognition and avoidance is experiential, I suggest you enroll in a weekend avalanche training course, even if you think you already know about avalanches. You will be surprised at the number of life-saving tips you can learn from a professional avalanche specialist.

Second, carry a shovel and wear an electronic rescue beacon. Make sure everyone in your party carries the same. Learn how to use them and practice, practice, practice. The only time I can think of not carrying this equipment would be in mid-summer, when a weather forecaster very confidently swears to the impossibility of any snow falling during your trip to the mountain glacier and the last snowfall was a month ago. These times are rare in the glaciated mountains of North America.

Third, snow stability can vary dramatically over short distances and short time spans. The most common causes of these variations are changes in terrain configuration, temperature, and snow distribution. For example, there is more stress on steep and convex slopes than shallow or concave slopes; the influence of temperature that can change snow structure will be unique for different elevations and different exposures to the sun; and the shallow, scoured snow cover on a windward slope will have a different type of instability than the wind-loaded snow pillows on leeward slopes.

Finally, snow avalanches are quite efficient at sweeping entire climbing parties to their death. In Alaska in 1988, four young men died on Mt. Foraker as they crossed a steep glaciated slope on their way to the summit. A slab avalanche less than 3 centimeters (1.2 in) thick broke away and swept them off their feet. Sliding on the exposed ice surface, they were unable to stop.

Another place for sure-fire disaster is above a crevasse. Avalanching snow can not only sweep you into the crevasse, but it will fill in on top of you, burying you, suffocating you, and wedging you tighter into the slot. I do not know of anyone who has survived this fate.

If you are caught in an avalanche, you must try to stop yourself from being swept away with the slide as soon as possible. Although most avalanche victims die of suffocation after being buried by a snow avalanche, the most common way that mountain climbers die in avalanches is by injuries received during the fall. Many are carried thousands of feet down a mountain. Others are swept into crevasses and crushed by the ensuing load of ice and snow.

If on skis, try to ski off to the side or toward something you can hold onto like a rock outcrop. If on foot, dig in with everything you have. The force of an avalanche is sudden and strong, often coming in waves and lasting for several seconds. Therefore, once you think the avalanche is over, keep your hold for a little while longer just to make sure.

Although there are some advantages to using a belay rope in avalanche terrain, special caution is advised. The force of an avalanche not only can pull the person that is in the middle of the slope away, but it can also pull the belayer. Then too, the force of the slide may break the rope entirely.

Using a rope that is too short can expose more than one person at a time to an avalanche. This can result in needless multiple burials.

If there is a threat of avalanches and you still want to have the security of a rope, in order to have only one person on a slope at a time you usually will have to lengthen the distance between rope partners. The rope should be as long as is necessary for one person to be anchored in a safe location while the other moves to another safe location (see Fig. 9.25). This may mean tying two ropes together (see Fig. 9.26).

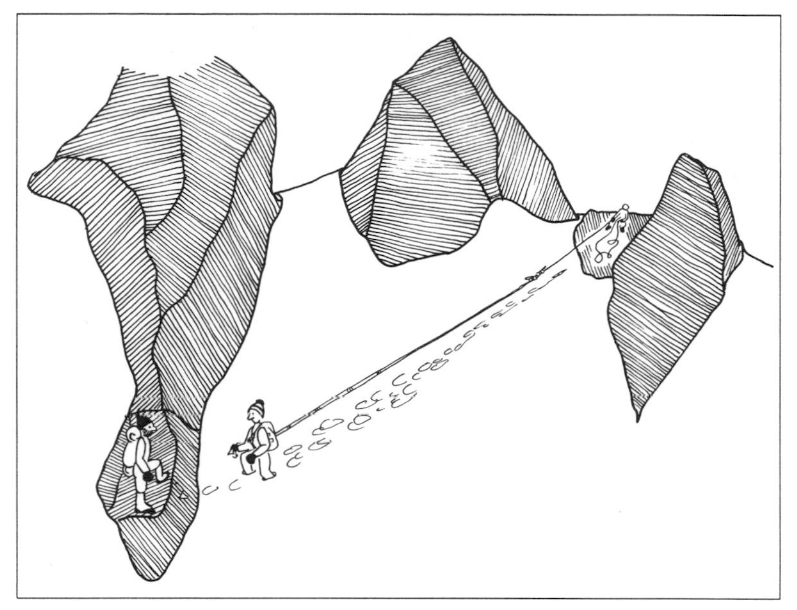

FIG. 9.25—Longer lengths of rope may be needed to belay through avalanche terrain. Travel from safe spot to safe spot to avoid both belayer and belayee from being swept away by the same avalanche.

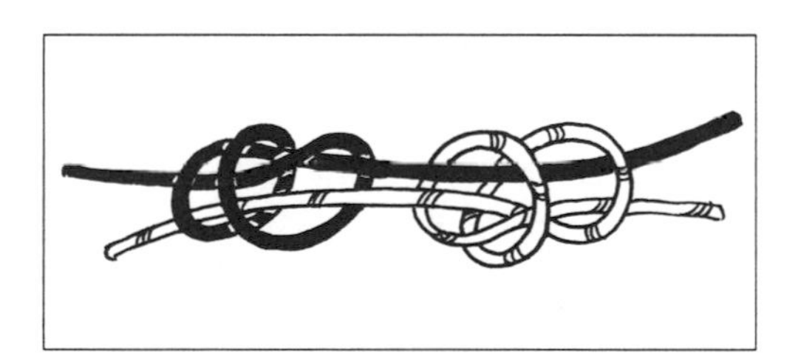

FIG. 9.26—A double fisherman's knot ties two ropes together. Loop the end of the first rope over the second. Come back and loop again, then thread the end through both loops. Do the same for the end of second rope over the first.

Ice Avalanches

Like snow avalanches, ice avalanches can happen at any time of year or any time of day. As you learned in Chapter Six they are caused by a combination of glacier movement, temperature, and serac configuration. Without a plethora of modern instruments (some of which have not even been invented) ice avalanches cannot be predicted.

Ice avalanches fall from hanging glaciers, icefalls, and any serac-covered portion of a glacier. Therefore, you must know the terrain above you. Many times the shape of the terrain or cloud cover will obscure your view. A topographic map will help.

If possible you would do well to simply avoid ice avalanche terrain altogether. This means that you should not walk up steep-walled glaciated valleys. Also, do not walk under obvious ice overhangs or unknown terrain. Glaciated terrain is sparse below latitude 45°N and there is only sporadic threat from ice avalanching. However, north of 45°N many mountain valleys are littered with glaciers and ice overhangs, and one would do well to pay attention at all times.

If you must pass beneath a region of seracs, pass quickly and one at a time if possible. Eleven people were killed by an ice avalanche on Mt. Rainier in 1981 while waiting for someone to check the route ahead.

Do not camp underneath ice avalanche terrain. I am not kidding; people do this and some die. It makes sense that the farther away from a serac-covered region the better. Ice avalanches have been known to travel miles down glacier, often across and over to the other side.

You are never completely safe in any valley that has a number of hanging glaciers on either side. If one side is less glaciated (and less snow-covered) than the other side, then edge your camp over to that side. If there is no clear choice, place your tent in the middle. If possible put it opposite rock ridges that may be between area seracs.

Even though ice avalanches can happen during any season and at any time of day, there are some critical times to especially avoid, like during surging events or times of thaw. However, these precautions will not protect against ice avalanches caused by spontaneous events like earthquakes, sonic booms, unexpected changes in temperature, or sudden glacial movements, etc. Simply said, there are major risks from ice avalanching, and it is important to be on guard at all times.

The Eye of the Cloud

Clouds and snow look the same when they are close together. When clouds surround the mountains whiteout conditions occur. This means that you cannot tell where the sky stops and the ground begins. It is easy to suffer vertigo. It is even easier to become lost.

Sometimes it helps to just stay put in whiteout conditions. Do not move. Put on some extra clothing to stay warm. Often visibility will return within a few minutes, especially in the spring or summer. You can afford to wait. If I were you I would only proceed if I were very, very certain that there was a smooth, uncrevassed way to a rock outcrop or snow-free glacier edge, and I could see some orienting features.

If you are caught in a snowstorm and are on snow-covered terrain, dig a hole or small cave. It is surprisingly warm underneath the snow. The air spaces between snowflakes act like a down blanket causing the snow cover to be a good insulator. This will also help you stay out of the wind.

In 1986, seven school children died partly because of whiteout conditions on a Mt. Hood glacier in Oregon. They stopped moving when they could no longer see where they were going and were disoriented. Unfortunately, they were not prepared for the ordeal and had not brought enough clothing even though they had managed to dig a shallow snow cave. They all died of hypothermia.

If you are an intrepid explorer who wants to continue moving, you must make certain that you (a) are headed in the correct direction and (b) can avoid hazards. Pull out your map, compass, and altimeter.* Know where you are now and where you want to end up. If you are smart you will place brightly colored sticks, or wands, into the glacier surface on your way up. If all else fails, you can at least find your way home by following these wands (usually spaced about 25 to 40 meters [80 to 130 ft] apart).

As you walk, poke your ice axe into the glacier surface ahead of you often, with every step if necessary. Throwing snowballs or tossing a length of rope in front of you can also help to "see" the terrain ahead. You are not only checking to feel where the glacier surface *is*, but you are also checking to make sure that you do not inadvertently walk over a cliff or into a crevasse.

Another signal for danger is a change in the surface texture. The change from relatively yielding snow or firn to hard, slippery ice could cause you to lose your footing and slip down the mountain. Changes in snow texture also may indicate a change in stability and the chance for avalanching.

Use a rope, especially if you have a long way to travel through this whiteout. Even if you do not expect to encounter dangerous terrain, you will be surprised at what can be discovered through the "eye of the cloud."

If you are on skis, you can get into trouble a whole lot faster. Do not be afraid to reduce your downhill style to a snowplow if you are having trouble seeing. Pride is not a consideration if your life is in danger. With luck you may be able to find a medial moraine or similar rocky feature to follow down glacier in a relatively crevasse-free region. Although cumbersome, you may want to rope up for better protection, especially if you are unfamiliar with or concerned about the terrain ahead.

Burning Desire

The sun can burn and it can blind. Simple facts but so many forget. Even if you are dark-skinned and need no sun protection elsewhere, I guarantee that sun exposure on a glacier will be quite uncomfortable without it. It is the ultimate tanning (burning) salon.

Firn and snow reflect 40 to 95 percent of the incoming solar radiation (see Fig. 9.27). This means you can burn from the bottom as well as the top. Think of the pain of sunburn inside the nostrils of your

*Use caution when reading an altimeter. Increasing clouds may signify that a storm is approaching with an associated center of low pressure. The lower atmospheric pressure will cause your altimeter to falsely indicate that you are at a higher elevation than is true.

nose, on the roof of your mouth, up your pant legs, in your arm pits, on that little soft spot under your chin, and on the tips of your ears.

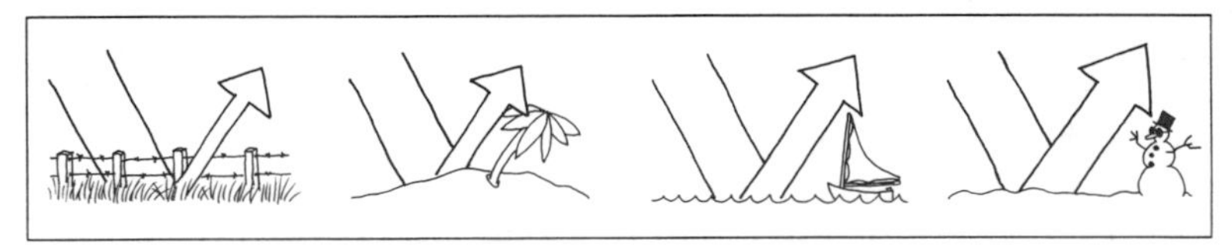

FIG. 9.27—Up to 25 percent of incoming solar radiation is reflected by grass, 30 to 40 percent by desert sand, 50 to 70 percent by water, and 70 to 95 percent by snow. (Firn reflects 40 to 60 percent and glacier ice reflects 20 to 40 percent of incoming solar radiation.)

If you do nothing else to protect yourself from the sun, be sure to wear sunglasses. These should be very dark with 100 percent UV protection. To prevent dangerous radiation from entering at any angle make or buy glasses equipped with side flaps. Tie a lasso around your neck and glasses so you will not lose them while peering down a crevasse. Some people like to tie their glasses tight to their head so they will not even think of falling off. Bring an extra pair just in case you break or lose yours or a friend forgets his or hers.

Those who have had the painful experience of snow blindness will never, ever forget their sunglasses. Even with cloud cover or fog, bright light reflecting from the snow surface of a glacier can burn and blind your vision. The agony of recovery is well remembered.

I know a wild guy who taught me exactly where the human body can tan if given the chance. He spends the major part of each summer naked on glaciers. The only tan line he has is around his eyes where his sunglasses reside. (I think he also has a tan line marking the location of his boots, but I have been too shy to let my eyes wander that far down his body to find out for sure.)

If you do not spend much time in the sun, then in addition to sunglasses, you will want to wear long pants, a long-sleeved shirt, and gloves. A brimmed hat that will cover your ears, forehead, and the back of your neck is also recommended. If it is hot outside just keep the clothing of light material and light colored or white, then it will be nice and comfortable. Put at least 29 SPF sunscreen on the exposed parts of your face and neck. A heavy coat of Desonex® or opaque, white diaper-rash cream is especially good for your nose, lips, and top of your cheeks. Test all skin creams before your trip to check for allergic reactions. Once in the sun, remember to reapply your protection cream regularly.

Even for those with plenty of previous sun exposure, it is still wise to use sunscreen on all exposed skin in addition to wearing sunglasses. Be aware that perspiration can cause sunscreen to drip into your eyes and irritate them severely enough to inhibit vision. Since this is the case you may want to wear a hat to shade your forehead instead of applying sunscreen there.

Outburst Floods

The sudden release of water from a glacier's plumbing system can cause outburst floods. Because the most dangerous floods seem to occur periodically, you can often learn of the potential for flooding by talking with local climbers or park and forest rangers. If you do hear the low thunder associated with rushing flood water, do not try to outrun it. Move to high ground immediately.

Ice Caves

Glacier ice caves are very beautiful, but because they are part of the glacier's transient plumbing system they can be dangerous. The best time to explore caves is in the fall or early winter, after the melt season has ended and before new snow can cover the entrance. Try to avoid spring and summer when rushing water can sweep you off your feet or when outburst floods are possible.

Ice caves are most accessible on stagnant or retreating glaciers. For this reason the roof of a cave entrance near the glacier margins may thin as the glacier surface melts away. A thin roof could collapse without warning. Also, rocks can slide off the slopes above entrances. Inspect all caves before entering them.

Roll Me Over

Sea kayaking is becoming such a popular way to observe tidal glaciers these days that a special word is needed about icebergs. I doubt that these small craft would have the same trouble as the Titanic, but the potential hazards are no less dangerous.

Blocks of ice fall off the end of a tidal glacier frequently. These plunging masses can create waves that are often quite large. Waves have been known to shipwreck fishing vessels and damage docks. Stay far enough away from the face of a tidal glacier so you will not be impacted by the falling ice, and so you can prepare for the subsequent wave action. Turn your vessel into approaching small waves to prevent rolling over sideways. For large waves you had better paddle away to avoid a perilous backward surfing experience.

As we learned in Chapter Four, icebergs periodically roll over and this motion can tip you over. Although this is more common with those bergs that look like camels with their head underwater, just about any shape has the potential for rolling. As you explore the icy wonder of a glacier fjord, view each iceberg from a respectable distance.

Rock Slides

The terrain around an active glacier may be newly sculpted by the moving ice. This means that it is often unstable. Always be alert for falling rocks. When walking on talus, try to step on the largest rocks. Watch to see how they are resting on the slope before transferring your whole weight. If the rocks do begin to move, they could entrain others surrounding you. To avoid being caught in a potential rock avalanche, keep moving toward the side, again trying to step on the largest available rocks.

Rocks lying over ice can be particularly dangerous. Obviously your feet slip away if you miss a rock and land on ice. Also, melting ice will release its hold on rocks, causing an entire rocky slope to become unstable, especially during the heat of the day.

Other Hazards

There are a variety of other potential dangers that you may encounter while venturing into the mountains to visit its glaciers. These include becoming lost, separating from your partners, battling inclement weather, and getting trapped by terrain that allowed access on the way in but not on the way out. The medical emergencies that you may encounter include high altitude pulmonary edema (HAPE), frostbite, hypothermia, and injuries from falling.

Before you venture too far into glaciated terrain, learn about additional emergency techniques needed to save your life and the lives of your exploring partners. The reference list at the end of this book includes some literature on mountain hazards, safe travel, and first aid. Also, there are many schools and training programs that can help. Check a local retail climbing store or mountaineering club to find out what courses may be offered.

Condition My Condition

The more physically fit you are, the more comfortable you will be while wandering around on a glacier. You want to be strong enough to feel unburdened by your hefty pack. You also want to have good leg strength to help you kick steps and climb steep slopes, good upper body strength to help pull yourself or a friend out of trouble, and good lungs to keep you going to the fascinating heights of brilliant névé-névé land.

Although most people are pretty well tuned into body fitness these days, let me share just a couple of ideas that have helped me prepare for glacier travel.

Because I ripped out the cartilage of one knee while doing the limbo dance at a Canadian ski lodge and did the same to the other knee while twisting to Chuck Berry's rock-n-roll in a tight, black satin skirt,

I pay particular attention to strengthening my thigh muscles, or quadriceps. This helps to hold my knees together while pounding down a mountain with stiff boots and a heavy pack. (Using two adjustable ski poles for support on the longer descents is a great aid for us bums with bad knees.)

I also do a number of sit-ups each day to strengthen my stomach muscles. This seems to reduce the strain on my back caused by the pack load.

Running and/or hopping up stairs helps the old legs stay in shape and contributes to aerobic conditioning. We have a nearby stadium with more stairs than one can imagine and some very handsome track stars to watch as well.

Upper body strength seems to come naturally for some men, but not so for women. I finally installed a chin-up bar in the door of my kitchen so I could maintain at least the ability to haul my own weight. If you find chin-ups difficult, or impossible, just hang from the bar for a while. This surprisingly increases arm strength quite well. After a few days of hanging around, several minutes at a time, begin pulling yourself up. Start with your hands wrapped around both sides of the bar and lift your chin to alternating sides. This will help increase forearm and shoulder strength simultaneously. With ten to twenty chin-ups each day, you can be confident of hauling yourself out of a crevasse or pulling a friend to safety.

So, to sum up, pay attention, think, be prepared for the worst, and have fun!

CHAPTER
—10—

Susie's Safe Travel Kit

The following is a list of basic equipment needed to travel safely on a glacier. There are many alternatives to these suggested items, and there may be some additional equipment you will want to match your own skills and needs. Use this as a guide, then check a local climbing store to see what else is available or what is new.

There are a couple of things to remember about equipment. One is, never leave your gear behind. Conditions can change quickly in the mountains. What you think you can do without in the morning may become so important in the afternoon that you would sell your soul for it. Also, what appears to be a short distance away from camp during sunny weather can turn into the long journey from hell if a sudden storm develops or an accident occurs.

Another thing about gear is that animals like to chew on it. This is especially true of leather that has absorbed salt, like boots and crampon straps. I am not referring to ice worms, but mountain goats, bear, deer, elk, moose, and rodents that wander on and around many of North America's glaciers.

Ice Axe

For general glacier travel the curved and serrated picks with flat or slightly curved adzes work fine. There are many kinds of materials. Rubber coatings on metal shafts make them warmer to carry but more difficult to push into the snow for anchoring. The new glass materials are as strong as metal and tend to be warmer. If you have an old wooden axe, it is probably weak and brittle. You are better off hanging it up on the wall and buying a new one.

Many glacier travellers who confine themselves to gentle slopes prefer shaft lengths of 70 to 90 centimeters (30 to 35 in). Those who seek better control on steep ice carry 30 to 50 centimeter (12 to 20 in) axes.

Ice axe picks come with *positive* clearance or *negative* clearance. Either will grip well in snow or firn, provided you have perfected your "ohmygodihopeicanstop" technique. A positive clearance allows better grip on ice (see Fig. 10.1a). However, if not handled properly, the gripping power of a positive clearance pick can cause the ice axe to fling out of control, rendering it completely useless for the sliding victim. This is why most glacier travellers prefer negative clearance picks (see Fig. 10.1b). Although with a negative clearance pick one may not be able to completely stop on ice, at least you should be able to hold on long enough to slow down and regain control.

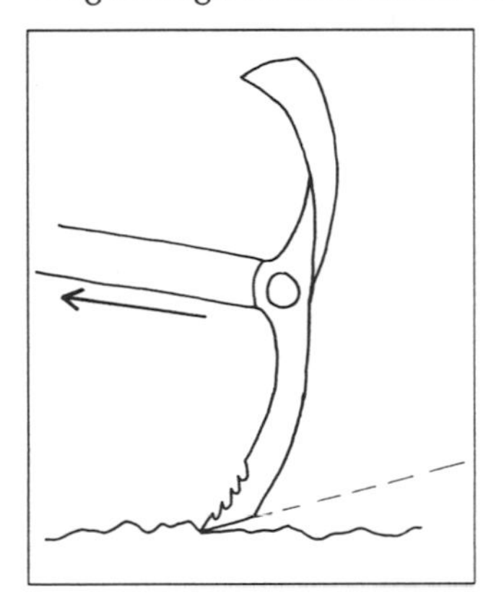

FIG. 10.1a—A positive clearance pick is best for gripping ice.

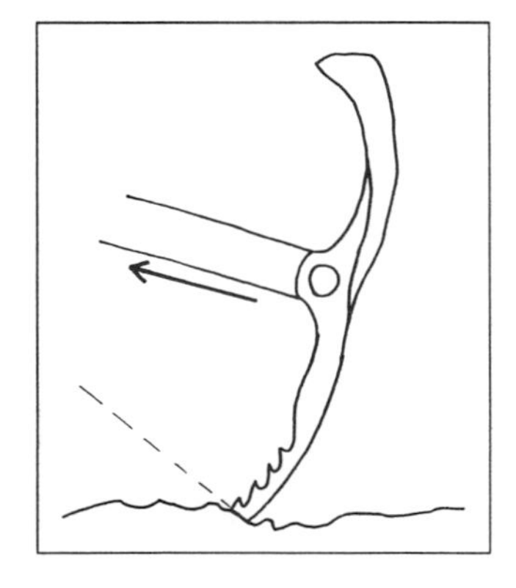

FIG. 10.1b—A negative clearance pick does not grip as well on ice but is easier to handle while sliding than a positive clearance pick. It is preferred by many glacier travellers.

There are some ski pole self-arrest grips available. Those with metal picks are best. Some work reasonably well on hard snow and firn. However, none are effective on ice. By skiing without your pole straps, or with straps that can slide down the shaft of your pole, you can slip yourself to the pole tip, forcing it into the ice with the weight of your upper torso. Hopefully, you can slow down enough to regain control on your skis.

Even great climbers have been known to drop their axes and look pretty silly. Use a wrist loop, or better yet, rig a long cord attached to your chest harness or a sling around your shoulders. This way the axe can be transferred easily from hand to hand. Remember that it becomes a dangerous weapon while tied to your body if you lose your grip during a fall. Keep a firm hand on your axe at all times.

Rope

Rope material is important. Primarily, it must be strong. The amount of stretch and elasticity you want in a rope depends upon the types of use you expect from it. For general glacier travel nylon kernmantel ropes are excellent.

Although a standard climbing rope is 11 millimeters in diameter, many glacier travellers prefer the smaller 9 millimeter ropes. They are lighter to carry. With two lengths of 9 millimeter instead of one length of 11 millimeter, you have much more versatility. If everyone in your party carries his or her own 25 meter (80 ft) section of rope, then you can distribute the weight and have plenty of rope for a variety of purposes.

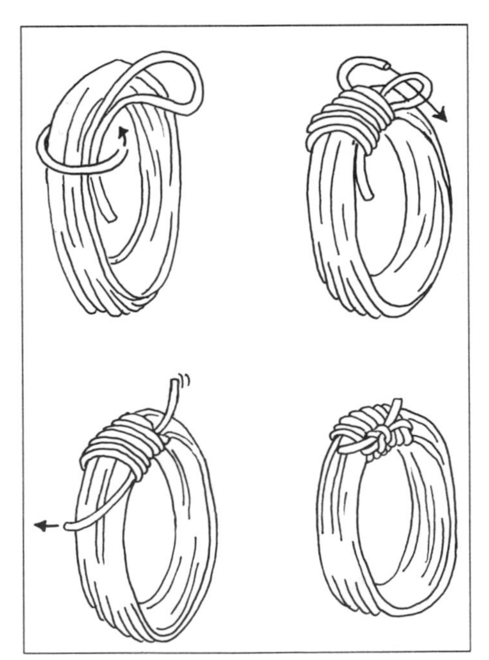

FIG. 10.2—Carry and store your rope in a way that it will not kink or become tangled. If carried in a coil, whip the ends to hold the coil together.

Because a rope can make the difference between life and death, taking care of it is important. Inspect it often for cuts and abrasions. If a strand is broken or if several strands are damaged, retire your rope to the practice barn or rope swing. Never walk on your rope. If it becomes dirty, wash it by hand with mild soap in lukewarm water. Do not use it for anything but glacier travel and climbing. Carry it and store it in an organized coil so it will not kink or tangle (see Fig. 10.2). Let it air dry before you store it in a cool, dark place. Keep it away from intense heat or chemicals that could deteriorate the nylon.

Prusiks

The prusik slings that are recommended in this book require two lengths of cord. One length of cord should be about 5 meters (16 ft) long to form a loop that can stretch from your foot to your shoulder. The other length of cord should be about 3 meters (10 ft) long to form a loop that can stretch from your seat to your shoulders. To form the loop, tie the loose ends together with a double fisherman's knot.

The diameter of the prusik cord should be thin enough to grip the rope but thick enough to be loosened easily. Usually a 5.5 millimeter diameter cord will work well on a 9 millimeter rope, whereas a 6 to 7 millimeter diameter cord will work on an 11 millimeter rope.

Harnesses

Chest and seat harnesses are relatively inexpensive these days. Even so, I usually bring a manufactured seat harness then carry extra 25 millimeter (1 in) webbing to make into a chest harness for moderate glacier travel. This cuts back on a little weight and gives me some flexibility to use the extra webbing for other purposes.

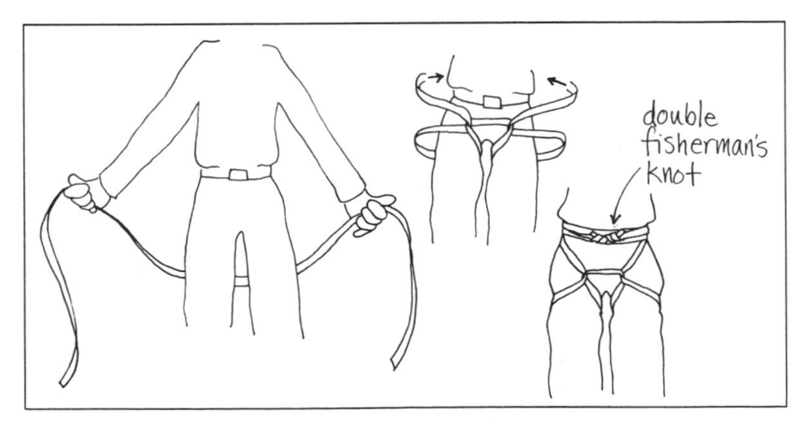

FIG. 10.3—A seat harness can be improvised with about 4.5 meters (15 ft) of 25 millimeter (1 in) webbing. Use a double fisherman's knot to tie the ends together.

Although there are several ways to improvise a seat harness, here is one that is simple, effective, and relatively comfortable for both men and women (see Fig. 10.3). Use about 4.5 meters (15 ft) of 25 millimeter (1 in) webbing. Stretch it behind your back, then pull a loop between your legs. Thread each end around your legs, just under your buttocks, through the crotch loop and back around your waist several times. Tie the ends together with a double fisherman's knot. You can improve this harness by making leg loops with overhand knots.

A simple chest harness can be improvised with about 2.5 meters (8 ft) of 25 millimeter (1 in) webbing (see Fig. 10.4). Tie the ends together with a double fisherman's knot to form a loop. Cross the loop behind your back, slip your arms through, then attach the two sides together over your chest with a locking carabiner, its gate facing out.

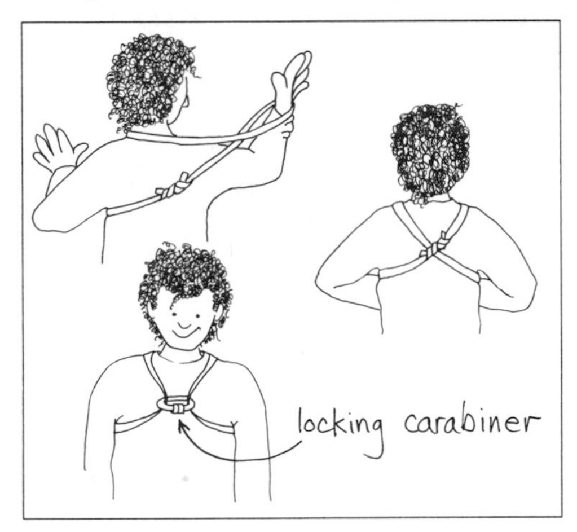

FIG. 10.4—A chest harness can be improvised with about 2.5 meters (8 ft) of 25 millimeter (1 in) webbing. Form a loop with a double fisherman's knot, then use a locking carabiner to connect the shoulders.

Remember that all harnesses need to be adjusted to fit comfortably, but snuggly, over your outer clothing. Tie into both seat and chest harnesses before you need them. Whether you make your own or rely on manufactured harnesses, take good care of them and frequently inspect for any fraying of the material or seams.

Crampons

Twelve-point crampons are the standard these days. They have two points in the front for front-pointing up steep ice pitches. With the front points you have to be careful not to trip.

Although using leather straps to attach crampons to your boots is still popular, they are becoming a major source of accidents as they break or are improperly attached. There are some quick release mechanisms that are now vogue and seem to be much safer. Either way, be sure you adjust the size and can secure crampons effectively before you need them. Carry an extra bit of wire or extra straps in case one of your crampon straps breaks.

Many skiers use crampons especially made to slip over the ski, under the boot. These types of crampons are called harscheisens and are very useful for steep traverses up hard snow.

Boots

The stiffer the sole of boot you have, the easier it will be to kick steps in hard snow and firn and to front-point with crampons. The softer the sole of boot you have, the easier it will be to walk without gathering a plethora of blisters. There are several boots available today that offer a compromise. You should choose one that will be best for your own style.

When buying boots, it is a good idea to try them out at home for a couple of days before you commit the wad of money. Make sure they are not too tight. Tight boots will restrict your blood flow and increase the likelihood of cold feet and possible frostbite. Do not buy boots that are too big either. A loose boot will foster blisters.

Leather boots that become wet are cold no matter what. They will conduct heat away from your foot. For this reason the leather and seams need to be regularly waterproofed. There are some new synthetic materials that are gentler on the leather and thread than the old boot greases and waxes, and they are as good or better at waterproofing. You can buy all this stuff where you buy the boot, and they will show you how to use it.

Many climbers now prefer plastic boots. Be sure to find a high-quality plastic boot with a good reputation. In extreme cold, some plastic boots have been known to crack.

Socks

Wear one thin sock next to your foot and a thick sock over that to help prevent blisters. Polypropylene is a wonder material that has the same warmth and moisture-wicking power of wool with less weight and scratchiness. But do not be bashful about wearing a wool sock. Wool still holds its own in the competitive sock world.

In very wet climates, I have taken to wrapping a plastic bag over my socks. This helps to keep my system dry no matter how wet the weather. By placing the plastic bag over your thin sock, you also can help keep dry in the new plastic boots by confining perspiration to the layer next to your foot and not letting it invade your whole system.

Gaiters

These handy little items keep snow and other forms of water from oozing down through the top of your boot. They now make them out of waterproof, breathable materials, which are good for warm weather.

Find a gaiter that you can put on and take off easily, especially if you spend the night in the mountains. I find the side entry, snap and Velcro® ones best for my liking. Make sure the Velcro® is totally closed because it does not work if it is coated with snow and ice. Zippers always break or freeze up on me whenever I am struggling with gear in a driving snowstorm.

I also like the full-length gaiters that fit over the top of my boot and clear up to my knee. There are some anklet varieties available that you might find fun to have in nice weather.

Insulated gaiters are made for extreme conditions. They not only cover your ankle and calf, but your entire foot and toe as well. Although quite toasty, they may be a bit much for many casual glacier activities.

You might carry an extra bit of string in case your gaiter strap breaks. There is nothing worse than having your gaiter ride up your leg and fill up with snow as you struggle along your way.

Carabiners

Carabiners are oval or D-shaped doo-dads that help you hold everything together. Locking 'biners have a screw type gate that you can secure shut (see Fig. 10.5a). These are necessary on your harness and for attaching to a protective rope. Less expensive 'biners have a snap-like gate and are handy for other stuff, like belaying packs, hanging laundry, etc. (see Fig. 10.5b). Carabiners with large gates are easier to use with gloved hands but may sacrifice some strength.

Most carabiners are now lightweight and sufficiently strong. However, it is best to buy only those that have been approved by the UIAA (Union Internationale des Associations d'Alpinisme). I carry three locking 'biners and three snap 'biners, and it seems to be enough for most situations.

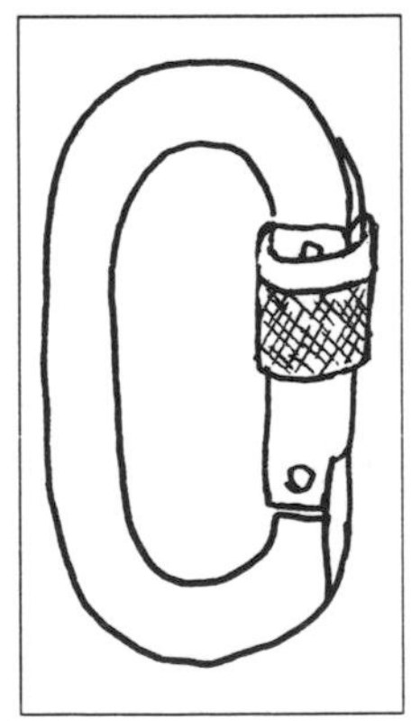

FIG. 10.5a—Locking carabiners should be used whenever securing a harness or attaching to a protective rope.

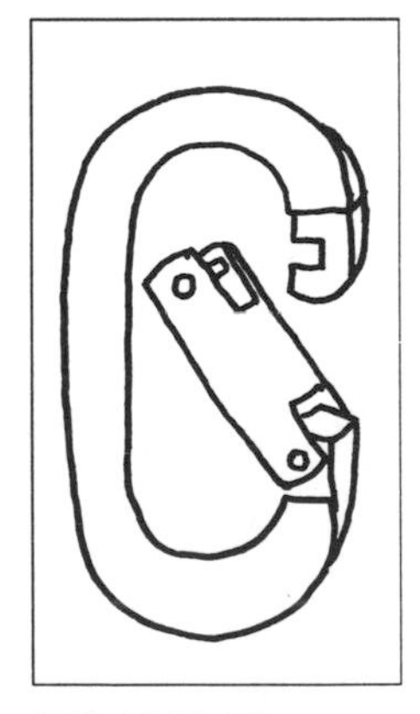

FIG. 10.5b—Snap carabiners can be used for a variety of purposes.

Avalanche Beacons

Avalanche beacons are high-frequency transceivers that allow people to find friends who may be buried in the snow. These are short-range beacons and work best if you begin a methodical search from the point where you last saw your friend.

Whenever you buy an avalanche beacon it will come with instructions on how to search most efficiently. However, it takes a lot of practice to be confident. When you bring a beacon, be sure to wear it next to your body, not in your pack.

Beacons with variable audio control and earphones are the easiest to use. However, delicate phone cords are fragile and break easily with extensive use. Some now have visual signals for the hearing impaired or when environmental noise interferes.

Check that everyone is carrying a beacon with the same frequency. The standard North American frequency is still 2,276 Hz. The standard European frequency is 457 KHz. The European frequency is now becoming the world standard, but in this period of transition many people have died because of incompatible frequencies. There are several dual-frequency beacons now available that are especially useful for those who travel to other countries.

Before you rely on your beacon, make sure it is in working order and has fresh batteries or is freshly recharged. Take out the batteries if your beacon is idle for long.

Avalanche beacons do go bad. Some transmit but do not receive. Others receive but do not transmit. Take care of your beacon in the same way you would any high performance machine.

Shovel

Although many day trippers and summertime travellers do not bother with shovels, it is surprising how many uses you can find for a good scooper. Your shovel should be strong enough to dig through hard avalanche debris, yet light and small enough so you will always want to carry it with you. Shovels also come in handy for digging snow caves, tanning solariums, and water catchment basins. More importantly, a shovel can be used to help investigate snow stability and build strong anchors for belaying and crevasse rescue.

Although there are many plastic shovels available, I prefer a good metal, like reinforced aluminum, for my shovel. Metal is an efficient digger because it has less give than plastic. D-shaped grips on the shovel handle work better than T-shapes or no grips at all, especially when wearing gloves. Many shovels are now made especially for mountaineering and are strong, lightweight, and small enough to fit inside a day pack.

PACK

There are many opinions about what kind of pack one should have. Internal frames are now the rage and for good reason. They are comfortable and allow you to tie the load close to your body to minimize any shifting of weight that may throw you off balance. Many packs now come with compression straps that allow you to change the size of your pack to fit its load. These are a great help in preventing weight shifts.

Some people like the convenience of zippered compartments on a pack. Be aware that zippers freeze in snowstorms, even those big plastic ones. Although one or two zippered compartments are nice, I try to find packs with big interiors and drawstring closures. Then I put pull tabs on all zippers so I can keep my gloves on when I work them.

Many people like to tie things on the outside of their pack like ice axes, skis, shovels, crampons, dirty sneakers, etc. All items need to be tied tight to the pack so they will not flop around and unexpectedly shift balance. When bushwhacking, junk on the outside of a pack can easily snag, so keep it to a minimum.

Ice axe loops are nearly standard on packs now. Straps or spaces between the side pockets to carry skis are also becoming standard.

So how big of a pack should you carry? Usually 55 to 75 liter (15 to 20 gal) capacity packs are sufficient for short trips to a friendly glacier. But you should figure it out for yourself. Here is a summary of the items you may want to carry.

altimeter
avalanche beacon
bivouac sack
brimmed sun hat
carabiners (three locking and three gate)
chest harness or about 5 meters (15 ft) of 25 millimeter (1 in) webbing
compass
crampons
first-aid kit
gaiters
gloves (one pair of lightweight inner, and one warm pair with leather palms)
headlamp
ice axe
insect repellent
insulating pad
knife
long pants
long-sleeved shirt
long underwear
map of area
matches (waterproof)
prusiks
rain gear
rope (25 meters [80 ft] of 9 millimeter)
seat harness
shorts
shovel
stocking hat
sunglasses (two pair)
sunscreen
water (2 liters [2 qt])
whistle

* * *

extra wire
extra cord, webbing, and/or straps
extra socks
extra warm clothing
extra food
extra headlamp batteries
extra headlamp bulbs
small pair of pliers
screwdriver
needle
thread
safety pins

APPENDIX

North American Glaciers: Where Are They?

With several hundred thousand glaciers in North America, it is impossible to visit them all. Two of the most popular are the Mendenhall in Alaska and the Athabasca in Alberta. I do believe that accessibility is the key to their popularity. The Mendenhall is on a city bus route from the capital of Juneau, and the Athabasca supports buses that are driven right onto the ice!

Not all glaciers are as easy to reach as those two. As a matter of fact, the last time I was bushwhacking through head-high scrub alder and devil's club, the face of my friend (without whose help I would have *never* found my way home) popped back behind a large leaf and said, "I think you should write about glacier access." I quickly agreed with him as I gingerly tried to pull stickers out of my forearm.

The recent retreat of so many mountain glaciers has caused dense, fast-growing weeds and brush to grow in place of their once extended snouts. This is especially true in the rain forest regions of southern Alaska and along the coast ranges of British Columbia and Washington. In addition, the outflow streams from these glaciers can create quite a barrier in themselves.

Once a glacier has retreated above tree line, access is simplified because there is little if any vegetation to get in the way. However, many of these small, lingering glaciers are clinging onto very rugged and often unstable terrain. This makes them almost as difficult to approach as their vegetation-shrouded cousins.

All things considered, deciding which glaciers to include in this inventory was rather easy. I simply chose those that can be reached within a day's hike from your automobile. A few others that require a longer hike or a different mode of transportation also are included because of their special interest.

Although today's glaciers may seem small when compared to their once grand size during the last ice age, they are no less impressive. Over 75,000 square kilometers (20 million acres) of glacier ice is held in Alaska, enough to cover the entire area of New England. About 50,000 square kilometers (10 million acres) are held in Canada's western mountains. Another 150,000 square kilometers (40 million acres) of permanent ice resides on Baffin, Axel Heiberg, Ellesmere, and Devon islands in the Northwest Territories. A little less than 600 square kilometers (150,000 acres) of glacier ice can be found in the contiguous United States, with over 75 percent of that in Washington state alone.

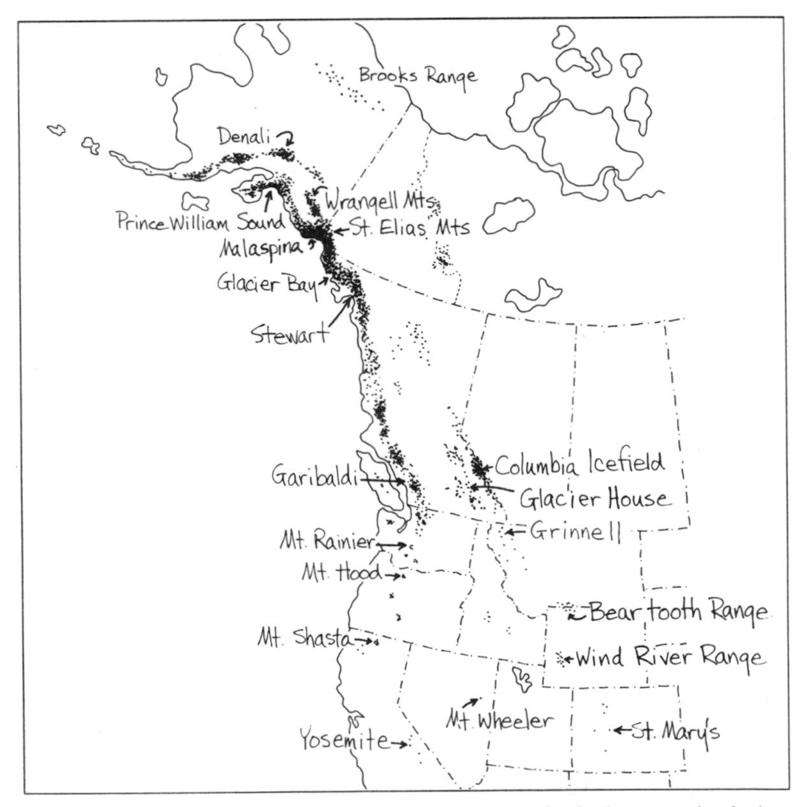

FIG. A.1—This map shows the approximate locations of today's mountain glaciers.

Most mountain glaciers in North America can be found in the St. Elias Range of Alaska and southwestern Yukon Territory (see Fig. A.1). The St. Elias and its neighboring Wrangell Mountains contain most of the surging glaciers found in this part of the world. Most of the tidewater glaciers flow from the Chugach, Fairweather, and St. Elias ranges.

Nearly all of North America's mountain glaciers are temperate. However, subpolar glaciers exist in the Brooks and Mackenzie ranges, and at high elevations in the St. Elias, Fairweather, and Alaska ranges. Polar glaciers lie in the northeast Canadian Arctic.

Many rock glaciers have been recognized in Colorado's Elk Mountains; the Blanco Massif; and the San Juan, Front, and Mosquito ranges. The Canadian Rockies, especially the Alberta side, and the Mackenzie Range of the Yukon Territory are littered with rock glaciers. The south slopes of Alaska's Wrangell Mountains and the Brooks and St. Elias ranges also support rock glaciers. Other rock glaciers have been found in the Absaroka and Yellowstone ranges of Wyoming, in the Table Cliffs Plateau and La Sal ranges of Utah, and in some of New Mexico's mountains.

The following list of glaciers includes much of their recognized variety, from grand piedmonts to remnant rock piles. A variety of geographic locations are also included. The glaciers are organized by country, state and/or province, and glaciated area, then alphabetical by glacier name. An address is provided that will help you obtain more information about the area or a particular glacier.

The approximate latitude and longitude are given for each glacier to help you find it. Also, the United States Geological Survey (USGS) or Energy, Mines, and Resources of Canada (EMRC) topographic map that covers its area is listed. In addition to these maps, many Canadian national and provincial parks and U.S. national forests and parks have maps available for their particular area. Contact each district office to find out the availability of such maps.

For more information about obtaining USGS maps, contact:

United States Geological Survey
Map Center
Branch of Distribution
Box 25286
Federal Center, Building 41
Denver, CO 80225
USA

For more information about obtaining EMRC maps, contact:

Canada Map Office
615 Booth Street
Ottawa, ON K1A 0E9
CANADA

The directions for reaching these glaciers assume summer travel. Many of the access roads and some of the visitor centers are closed during winter. However, it is not uncommon to enjoy glacier travel during the winter. In fact, some access routes are made easier when surrounding vegetation is covered with snow. Contact the local park or forest office for further information.

— CANADA —

Alberta

Jasper National Park

Superintendent
Box 10
Jasper, AB T0E 1E0
(403) 852-6161

COLUMBIA ICEFIELD

The Columbia Icefield is one of the largest in the North American Rockies. It is considered a hydrographic center of our continent. This means that it supplies water to all the major drainage systems: north to the Arctic, west to the Pacific, east to the Great Lakes and Atlantic, and south to the U.S. Columbia system. To obtain more information about the icefield, contact the Columbia Icefield Center, (403) 761-7030. It is open during the summer months.

ANGEL

Type: hanging cirque
Lat/Long: 52° 41' / 118° 04'
Map: 1:50,000 EMRC Amethyst Lakes 83D/9
Approx. Area: 0.9 sq km (220 acres)

Directions: The trail leaves from Mt. Edith Cavell parking lot in Jasper National Park. The glacier is on the north face of Mt. Edith Cavell and can be seen from the teahouse.

Notes: This is not only a spectacularly beautiful glacier, but it appears to have a rock glacier or glacier remanié at its base.

ATHABASCA

Type: valley
Lat/Long: 52° 12' / 117° 15'
Map: 1:50,000 EMRC Columbia Icefield 83C/3
Approx. Area: 15 sq km (3,700 acres)

Directions: This glacier is just opposite the Icefields Center on the Icefields Parkway. There are balloon-tire buses that drive onto the glacier and several options for guided tours.

Notes: The Athabasca is Canada's most popular glacier. It is also one of the most well studied. There are three spectacular icefalls, moulins, glacier caves, nearby hanging glaciers, and guided walks to keep visitors busy.

Banff National Park

Superintendent
Box 900
Banff, AB T0L 0C0
(403) 762-3324

HORESHOE

Type: remanié
Lat/Long: 51° 21' / 116° 16'
Map: 1:50,000 EMRC Lake Louise 82N/8
Approx. Area: 5 sq km (1,200 acres)

Directions: Follow the Paradise Creek Trail from Chateau Lake Louise to the Giant Steps.

Notes: I do not know if this glacier has been officially designated a remanié, but informal discussions about Horseshoe among scientists refer to it as a remanié. It begins at an elevation well below the perennial snow line and is at the base of steep cliffs whose avalanching snow and ice provide nourishment.

SASKATCHEWAN

Type: valley
Lat/Long: 52° 12' / 117° 08'
Map: 1:50,000 EMRC Columbia Icefield 83C/3
Approx. Area: 30 sq km (7,400 acres)

Directions: From the Icefields Parkway, just south of Icefields Center, follow the Parker Ridge Trail for a wonderful view of Saskatchewan, its medial moraine, and the Columbia Icefield. Or park near Bridal Veil Falls and walk up the North Saskatchewan River to the glacier's snout.

VICTORIA

Type: valley
Lat/Long: 51° 23' / 116° 17'
Map: 1:50,000 EMRC Lake Louise 82N/8
Approx. Area: 3.5 sq km (865 acres)

Directions: Although visible from Chateau Lake Louise, you will want to hike closer by following the lake shore trail. Be sure to stop for refreshments at each teahouse along the way.

Notes: Built in 1890, the Chateau Lake Louise provided an early base for amateur and professional scientists. The Victoria is just one of the many nearby glaciers that have a long history of study.

WENKCHEMNA

Type: rock (avalanche)
Lat/Long: 51° 19' / 116° 14'
Map: 1:50,000 EMRC Lake Louise 82N/8
Approx. Area: 4 sq km (990 acres)

Directions: The trail leaves from Lake Louise Park Lodge through the valley of Ten Peaks to Moraine Lake Lodge. Follow the Eiffel Lake Trail along the north flank of the glacier for a spectacular view.

Notes: This glacier flows 90 degrees from adjacent couloirs that provide its source of ice, snow, and rock.

Kananaskis County

Peter Lougheed Provincial Park
Box 130
Kananaskis, AB TOL 2HO
(403) 591-7222

ROCK

Type: rock (moraine)
Lat/Long: 50° 48' / 115° 20'
Map: 1:50,000 Spray Lakes Reservoir 82J/14
Approx. Area: 1 sq km (250 acres)

Directions: From Canada Highway 1, exit at Kananaskis Country. Drive south on Highway 40 (closed in winter) for approximately 50 kilometers (30 miles). There is a sign and a short turnoff to the rock glacier.

Notes: The glacier ice buried underneath this moraine is thought to have originated during the extensive glaciation about ten thousand years ago.

British Columbia

Garibaldi Provincial Park

Garibaldi/Sunshine Coast District Office
Alice Lake Provincial Park
Box 220
Brackendale, BC VON 1HO
(604) 898-9313

HELM

Type: valley
Lat/Long: 49° 58' / 123° 00'
Map: 1:50,000 EMRC Cheakamus River 92G/14
Approx. Area: 3 sq km (740 acres)

Directions: From Canada Highway 99 turn east at Shadow Dam (approximately 5 kilometers [3 mi] south of Whistler Village). Park at Rubble Creek and follow the Garibaldi Lake Trail. From Garibaldi Lake hike along the Panorama Ridge Trail. As soon as you see the glacier you can cut cross-country to it or continue along the ridge for a spectacular view that includes Garibaldi Névé and the Warren Glacier.

Notes: Photo studies have been done on the Helm for a number of years and mass-balance observations occurred during the mid 1970s as an extension of work on the Sentinel Glacier. The Sentinel, on the other side of Garibaldi Lake, is one of the few glaciers with a long-term mass-balance record. Scientists reach that glacier by boat.

Blackcomb/Whistler Recreation Area

Whistler Area Information Center
Box 181
Whistler, BC V0N 1B0
(604) 932-5528

BLACKCOMB

Type: cirque
Lat/Long: 50° 10' / 122° 50'
Map: 1:50,000 EMRC Alta Lake 92J/2
Approx. Area: 3 sq km (740 acres)

Directions: Ski lifts access the glacier from Whistler Village.

Notes: This glacier is used extensively for skiing. The ability to ski during the summer makes it a favorite for budding young racers.

Bugaboo Glacier Provincial Park

Ministry of Parks
Regional Director
101, 1050 West Columbia St.
Kamloops, BC V2C 1L4
(604) 828-4501

BUGABOO

Type: valley
Lat/Long: 50° 44' / 116° 47'
Map: 1:50,000 EMRC Howser Creek 82K/10
Approx. Area: 5 sq km (1,200 acres)

Directions: From Brisco (27 kilometers [17 mi] north of Radium Hot Springs on Highway 95) travel west on the gravel road about 45 kilometers (28 mi) to the park's parking lot (road closed in winter). Take the trail toward Conrad Kain Hut that switches back up near Bugaboo's snout.

Notes: This area is popular for mountaineering and used for heli-skiing in winter and heli-hiking in summer.

Glacier National Park

Superintendent
P.O. Box 350
Revelstoke, BC V0E 2S0
(604) 837-5155

ILLECILLEWAET

Type: valley
Lat/Long: 51° 14' / 117° 27'
Map: 1:50,000 EMRC Mount Wheeler 82N/3
Approx. Area: 6.4 sq km (1,600 acres)

Directions: About 4 kilometers (2.5 mi) south of Rogers Pass on Canada's Highway 1, turn off to Illecillewaet campground. Follow the Glacier Crest Trail to the toe of the glacier.

Notes: The Glacier House was built originally as a place to eat for rail passengers because a restaurant car was too heavy to pull up the pass. Many eastern families would extend their stay here to study nearby glaciers. Their studies of the Illecillewaet ("Great Glacier") began in 1887.

Elk Lakes Provincial Park

Ministry of Parks
District Manager
Box 118
Wasa, BC V0B 2K0
(604) 422-3212

PÉTAIN

Type: cirque-valley
Lat/Long: 50° 32' / 115° 10'
Map: 1:50,000 EMRC Kananaskis Lakes 82J/11
Approx. Area: 1 sq km (250 acres)

Directions: At Sparwood, on Highway 3, turn north from Highway 43 for 35 kilometers (22 mi) to Elkford. From Elkford travel gravel road 87 kilometers (54 mi) to the park. From park headquarters follow the Pétain Creek Waterfall Trail.

Stewart Area

British Columbia Parks
Bag 5000
Smithers, BC V0J 2N0
(604) 847-7320

BEAR RIVER

Type: valley
Lat/Long: 56° 05' / 129° 41'
Map: 1:50,000 EMRC Bear River 104A/4
Approx. Area: 4 sq km (990 acres)

Directions: Drive approximately 30 kilometers (19 mi) northeast of Stewart on Highway 37A. The glacier terminates into Strohn Lake, which is next to the road on the south side at Bear River Pass.

Notes: There is a rest area nearby. Although not now part of public lands, the Bear River Glacier area is being considered for park or recreation status in the near future.

BERENDON

Type: valley
Lat/Long: 56° 15' / 130° 10'
Map: 1:50,000 EMRC Leduc Glacier 104B/1
1:50,000 EMRC Frank Mackie 104B/8
Approx. Area: 11 sq km (2,700 acres)

Directions: Drive approximately 40 kilometers (25 mi) north from Hyder (near Stewart) on the Granduc Mine road. The glacier terminates just north of the mining center.

Notes: The mine portal lies within old moraines of the Berendon and an access tunnel exists underneath this glacier and its larger neighbor, the Frank Mackie Glacier. (Not part of public lands.)

SALMON

Type: valley
Lat/Long: 56° 08' / 130° 04'
Map: 1:50,000 EMRC Leduc Glacier 104B/1 & B/2
Approx. Area: 16 sq km (4,000 acres)

Directions: Drive north from Hyden (near Stewart) on the Granduc Mine Road. The road follows the Salmon River and soon travels adjacent to the Salmon Glacier. (Not part of public lands.)

Notes: The glacier periodically blocks Summit Lake to the north. Subsequent jökulhlaups destroy the road.

Yoho National Park

P.O. Box 99
Field, BC V0A 1G0
(604) 343-6324

YOHO

Type: valley
Lat/Long: 51° 36' / 116° 33'
Map: 1:50,000 EMRC Blaeberry River 83N/10
Approx. Area: 4 sq km (990 acres)

Directions: Drive about 5 kilometers (3 mi) east of Field on Trans-Canada Highway 1. Turn north toward Takkakaw Falls on the Yoho Valley road and drive for approximately 15 kilometers (9 mi). The Yoho Glacier trail begins at the parking lot. Walk about 8 kilometers (5 mi) to the glacier snout.

Notes: This glacier was studied extensively during the early 1900s.

Northwest Territories

Auyuittuq National Park

Pangnirtung, NWT X0A 0R0

PENNY ICE CAP

Type: ice cap
Lat/Long: 67° 20' / 66° 00'
Map: 1:500,000 EMRC Nettilling Lake 26NW & 26NE
1:250,000 EMRC Pangnirtung 26I
Approx. Area: 6,000 sq km (1.5 million acres)

Notes: This park provides a unique opportunity to visit an ice cap and its associated polar and subpolar glaciers. Contact park headquarters for further information on how to reach the area and the best mode of travel. A road from Pangnirtung through the pass to North Pangnirtung Fjord goes past over a dozen valley glaciers, hanging glaciers, and icefields.

Yukon Territory

Wrangell-St. Elias

Superintendent
Kluane National Park and Preserve
Haines Junction, YT Y0B 1L0
(403) 634-2252

Map: 1:250,000 EMRC 115B & 115C

This area offers access to many major icefields, including the Bagley and Seward icefields that are the beginnings for the great Bering and Malaspina glaciers, respectively. In addition, many of the tidewater glaciers that flow into Glacier Bay National Park begin here. Access is mostly cross country so contact the park for further information.

— UNITED STATES —

Alaska

Anchorage Area

Department of Natural Resources
Division of Land and Water Management
Mat-su/ Copper Basin Area Office
1830 East Parks Highway
Wasilla, AK 99687
(907) 367-4595

KNIK

Type: valley, cause of periodic jökulhlaups
Lat/Long: 61° 20' / 148° 15'
Map: 1:250,000 USGS Anchorage
Approx. Area: 600 sq km (150,000 acres)

Directions: From the old Glenn Highway out of Anchorage, turn east onto Knik River Road just before you cross the Knik River. Drive approximately 16 kilometers (10 mi) to the end of the road. The terminus is approximately 16 kilometers (10 mi) cross country. A viewpoint can be reached within an hour. (There is a proposal to install a tramway or extend the Knik River Road another 3 kilometers [2 miles] to the overlook.)

Notes: The Knik Glacier used to extend across the valley and dam the outlet to Lake George each year. When the dam broke subsequent flooding could release over 700 billion liters (185 billion gal) of water in less than a day. The jökulhlaup was so regular that the Alaska railroad included costs to repair damages from the expected floods in each annual budget.

Copper River Delta

Chugach National Forest
Cordova Ranger District
P.O. Box 280
Cordova, AK 99574
(907) 424-7661

One way to see all of these glaciers is to drive the Copper River Highway from Cordova. Another way is to take a raft trip down the Copper River from Chitina.

CHILDS

Type: valley, calving
Lat/Long: 60° 37' / 144° 55'
Map: 1:63,360 USGS Cordova, C-3
Approx. Area: 100 sq km (25,000 acres)

Directions: 77 kilometers (48 mi) from Cordova along the Copper River Highway. A side road leads to the viewing area that looks west across the river.

Notes: Cliffs of ice tower 75 to 90 meters (250 to 300 ft). There is a story that people wait along the Copper River shoreline for the Childs to break loose an iceberg. Subsequent waves have been known to throw salmon onto the beach. In 1910 the Childs Glacier advanced within 440 meters (1440 ft) of the Million Dollar Bridge. Fortunately the advance occurred during summer when the Copper River was full of meltwater. The deep water caused extensive calving that held the advancing front from reaching the bridge.

MILES

Type: valley, calving, periodic jökulhlaups
Lat/Long: 60° 37' / 144° 15'
Map: 1:63,360 USGS Cordova, C-1 & C-2
Approx. Area: 400 sq km (10,000 acres)

Directions: 77 kilometers (48 mi) from Cordova along the Copper River Highway a viewpoint looks east from the Million Dollar Bridge.

Notes: In 1885 and 1888 the Miles was within 110 meters (360 ft) of the Million Dollar Bridge. Then in 1909 a jökulhlaup from the Miles nearly destroyed the Million Dollar Bridge by raising the river level at the bridge from 5.5 to 11 meters (20 to 40 ft) in six days. After surviving that flood, the bridge was ultimately destroyed in Alaska's 1964 earthquake.

SHERIDAN

Type: valley, calving
Lat/Long: 60° 37' / 145° 15'
Map: 1:63,360 USGS Cordova C-3 & C-4
Approx. Area: 100 sq km (25,000 acres)

Directions: 22 kilometers (14 mi) from Cordova, just past the airport, drive north onto Sheridan Glacier Road. A side road leads to the viewing area at glacier terminus.

Notes: The glacier is retreating. This allows relatively easy access onto the ice.

SHERMAN

Type: valley
Lat/Long: 60° 32' / 145° 10'
Map: 1:63,360 USGS Cordova C-3 & C-4
Approx. Area: 55 sq km (14,000 acres)

Directions: 22 kilometers (14 mi) from Cordova, just past the airport, drive north onto Sheridan Glacier Road. A side road leads to the viewing area. Look east across the Sheridan Glacier terminus.

Notes: A large landslide that was triggered by the 1964 Alaskan earthquake covers the lower half of Sherman and is helping to insulate it from ablation.

Denali National Park

Superintendent
Denali National Park and Preserve
P.O. Box 9
Denali Park, AK 99755
(907) 683-2686 (recorded information)

The Denali massif contains numerous glaciers that are visible from a distance. However, accessing the glaciers requires significant time and skill. Contact the park or local guide services about the best way to reach Denali's glaciers.

MULDROW

Type: valley, surging
Lat/Long: 63° 04' / 151° 00'
Map: 1:63,360 USGS Mt. McKinley B-1
Approx. Area: 520 sq km (130,000 acres)

Directions: The park entrance is 380 kilometers (236 mi) north of Anchorage on Alaska Highway 3. Check in at the Visitor Access Center, to the right shortly after you enter the park. Unfortunately, private vehicles are restricted beyond Savage River (24 kilometers [15 mi]) into the park. However, a shuttle bus operates regularly. Round-trip to the glacier's terminus takes approximately nine hours on the bus. A difficult one- to two-day hike is required to reach the glacier itself. If the shuttle bus does not suit you, there is an end-of-the-year lottery for vehicle traffic in the park, contact park headquarters for more information.

Notes: The Muldrow offers one of the closest roadside views of Denali's many glaciers. It's terminus now is nearly stagnant so it is dirt covered and appears dark and hummocky. There are trees and brush growing on top of the dirt-covered ice. However, in 1956–1957 the glacier surged and the thickness of its terminus increased 200 meters (660 ft).

Glacier Bay National Park and Preserve

Superintendent
Glacier Bay National Park and Preserve
Gustavus, AK 99826
(907) 697-2230

There are numerous glaciers, icefields, and about one dozen active tidewater glaciers in the park that flow from the St. Elias and Fairweather ranges. Some glaciers are advancing, others retreating; many have spectacular displays of calving. Several fjords are choked with icebergs.

No easy foot access is available unless arrangements are made to be dropped off by boat or plane. There are several tour boats that sail to upper Glacier Bay from Gustavus to visit calving tidewater glaciers. Kayakers, canoeists, and backpackers may use tour boats to access upper Glacier Bay. Scenic flights also leave from Gustavus and float plane taxi service is available to the backcountry (check first about transporting rigid boats).

GRAND PACIFIC

Type: valley, tidewater
Lat/Long: 59° 10' / 137° 15'
Map: 1:250,000 USGS Skagway
Approx. Area: 200 sq km (50,000 acres)

Notes: About thirty years ago this glacier retreated so far back into Canada that the Canadians considered building a seaport there. Since then it has readvanced into U.S. territory. Birds wait for ice to calve off the glacier so they can feed upon the shrimp and krill that are churned up by the turbulent water. Part of the Grand Pacific's source area flows into Canada to feed the Melbern Glacier.

JOHNS HOPKINS

Type: valley, tidewater
Lat/Long: 58° 48' / 137° 10'
Map: 1:250,000 USGS Mt. Fairweather
Approx. Area: 60 sq km (15,000 acres)

Notes: Meltwater conduits have been known to empty out from the wall of the glacier's terminus. This causes meltwater to spew out the glacier front much like a Greek fountain.

Glenn Highway

Alaska State Parks
Matanuska, Susitna, Valdez, and Copper River
HC 32, Box 6706
Wasilla, AK 99687
(907) 745-3975

MATANUSKA

Type: valley
Lat/Long: 61° 40' / 147° 30'
Map: 1:250,000 USGS Anchorage
Approx. Area: 320 sq km (80,000 acres)

Directions: Drive about 150 kilometers (90 mi) northeast of Anchorage along the Glenn Highway to Glacier Point. From there a gravel road leads to a parking lot. Access to the glacier terminus is over private land and a fee is usually required.

Notes: The terminus is debris covered. There are fascinating medial moraines, and skiing on the névé is wonderful.

Portage Valley

Chugach National Forest
Glacier Ranger District
Monarch Mine Rd.
P.O. Box 129
Girdwood, AK 99587
(907) 783-3242

The USDA Forest Service operates the Begich, Boggs Visitor Center daily during the summer months and on a limited basis at other times.

BYRON

Type: steep valley
Lat/Long: 61° 45' / 148° 52'
Map: 1:63,360 USGS Seward C-5 & D-5
Approx. Area: 3.5 sq km (865 acres)

Directions: The Portage Glacier turnoff is about 80 kilometers (50 mi) southeast of Anchorage along the Seward Highway. Follow the paved road about 8 kilometers (5 mi) to the Begich, Boggs Visitor Center.

Notes: The best viewpoint is from a pile of snow built by continuous avalanching that is near the end of a maintained trail. Look for ice worms on the snow cone, or join a Forest Service interpreter for an evening "iceworm safari." Because of numerous snow avalanches, this valley can be extremely dangerous during winter.

EXPLORER

Type: hanging
Lat/Long: 61° 47' / 148° 56'
Map: 1:63,360 USGS Seward D-6
Approx. Area: 3 sq km (740 acres)

Directions: The Portage Glacier turnoff is about 80 kilometers (50 mi) southeast of Anchorage along the Seward Highway. Follow the paved road about 4 kilometers (2.5 mi) to the Explorer Glacier vista, near Beaver Pond Campground.

Notes: Though small and retreating back from its cliff edge, the Explorer still offers excellent photo opportunities.

MIDDLE

Type: hanging
Lat/Long: 61° 46' / 148° 55'
Map: 1:63,360 USGS Seward D-6
Approx. Area: 4 sq km (990 acres)

Directions: The Portage Glacier turnoff is about 80 kilometers (50 mi) southeast of Anchorage along the Seward Highway. Follow the paved road about 6 kilometers (4 mi) to the Williwaw Campground.

Notes: Like the Explorer, the Middle Glacier has retreated a bit back from its cliff-ending terminus. However, it remains picturesque.

PORTAGE

Type: valley
Lat/Long: 61° 44' / 148° 50'
Map: 1:63,360 USGS Seward C-5 & D-5
Approx. Area: 30 sq km (7,400 acres)

Directions: The Portage Glacier turnoff is about 80 kilometers (50 mi) southeast of Anchorage along the Seward Highway. Follow the paved road about 8 kilometers (5 mi) to the Begich, Boggs Visitor Center.

Notes: Join a Forest Service interpreter for a boat tour in Portage Lake to the glacier terminus. Because the glacier has been retreating approximately 15 meters (50 ft) each year, it is estimated that it will be out of view of the visitor center by the year 2020.

Prince William Sound

Chugach National Forest
201 East 9th Avenue
Anchorage, AK 99501
(907) 271-2500

There are about twenty active tidewater glaciers in Prince William Sound that flow from Alaska's Chugach Range. Most have been slowly retreating during the last fifty years, except for the Harvard, Harriman, and Meares.

Glaciers flowing into Prince William Sound can be accessed by boat or an extensive overland trip. For help in reaching the glaciers there are guide services and air and boat excursions available from Whittier and Valdez.

COLUMBIA

Type: valley, tidewater
Lat/Long: 61° 10' / 147° 00'
Map: 1:250,000 USGS Valdez and Anchorage
Approx. Area: 1,100 sq km (270,000 acres)

Directions: This glacier flows into Columbia Bay. Its terminus can be seen by boat and the névé can be reached after extensive overland travel.

Notes: This glacier is expected to retreat for another twenty to thirty years, backing out of its long, deep fjord. Small icebergs (about half the size of a supertanker) continue to calve and drift into shipping lanes. Oil tankers leaving from Valdez have special radar to watch for bergs. The glacier is being extensively studied, mainly because of its impact on shipping lanes.

HARVARD

Type: valley, tidewater
Lat/Long: 61° 25' / 147° 30'
Map: 1:250,000 USGS Anchorage
Approx. Area: 500 sq km (120,000 acres)

Directions: This glacier flows into College Fjord. Its terminus can be seen by boat, and the névé can be reached after extensive overland travel.

Notes: This glacier is slowly pushing its moraine shoal forward and advancing, providing a stark contrast to its neighbor, the Yale Glacier.

SHOUP

Type: valley, tidewater
Lat/Long: 61° 15' / 146° 30'
Map: 1:250,000 USGS Valdez
Approx. Area: 150 sq km (37,000 acres)

Directions: This glacier flows into Port Valdez. Its terminus can be seen by boat, and the névé can be reached after extensive overland travel.

Notes: This glacier calves into Port Valdez and has an unusually sharp right-angle turn near its terminus.

VALDEZ

Type: valley, calving
Lat/Long: 61° 15' / 146° 15'
Map: 1:250,000 USGS Valdez
Approx. Area: 75 sq km (18,500 acres)

Directions: Drive about 12 kilometers (7.5 mi) northeast of Valdez toward Valdez airport. The end of the road and the glacier terminus are about 2 kilometers (1.2 mi) past the airport. View is across its terminal lake.

Notes: This glacier used to serve as a trail to the interior at the turn of the century.

YALE

Type: valley, tidewater
Lat/Long: 61° 17' / 147° 30'
Map: 1:250,000 USGS Anchorage
Approx. Area: 220 sq km (54,000 acres)

Directions: This glacier flows into College Fjord. Its terminus can be seen by boat and the névé can be reached after extensive overland travel.

Notes: This glacier has ablated behind its moraine shoal and is now in a catastrophic retreat, providing a stark contrast to its neighbor, the Harvard Glacier.

Richardson Highway

Bureau of Land Management
Glennallen District 0
Box 147
Glennallen, AK 99588
(907) 822-3217

BLACK RAPIDS

Type: valley, surging
Lat/Long: 63° 26' / 146° 00'
Map: 1:63,360 USGS Mt. Hayes
Approx. Area: 340 sq km (84,000 acres)

Directions: This glacier flows toward the Richardson Highway from the west, near Rapids. Access requires a dangerous ford across the Delta River.

Notes: In 1937 the Black Rapids surged toward the Rapids Road House, making national news as Mrs. H. E. Revell described the towering wall of ice outside her kitchen window. The glacier stopped within 240 meters (790 ft) of her establishment. Some believe that the Black Rapids is due to surge again. If it does so, it could threaten the Trans-Alaska Pipeline.

CANWELL

Type: valley
Lat/Long: 63° 20' / 145° 30'
Map: 1:63,360 USGS Mt. Hayes B-4 and B-5
Approx. Area: 80 sq km (20,000 acres)

Directions: This glacier flows toward the Richardson Highway from the east, just south of the Fels Glacier. Access is relatively straightforward from the highway.

FELS

Type: valley
Lat/Long: 63° 22' / 145° 30'
Map: 1:63,360 USGS Mt. Hayes B-5
Approx. Area: 15 sq km (3,700 acres)

Directions: This glacier flows toward the Richardson Highway from the east, between the Canwell and Castner glaciers. Access is relatively straightforward from the highway.

CASTNER

Type: valley
Lat/Long: 63° 25' / 145° 30'
Map: 1:63,360 USGS Mt. Hayes B-4 and B-5
Approx. Area: 40 sq km (10,000 acres)

Directions: This glacier flows toward the Richardson Highway from the east, just south of Rapids. Access is relatively straightforward from the highway.

Thompson Pass

Alaska State Parks
Matanuska, Susitna, Valdez, and Copper River
HC 32, Box 6706
Wasilla, AK 99687
(907) 745-3975

WORTHINGTON

Type: valley
Lat/Long: 61° 10' / 146° 45'
Map: 1:63,360 USGS Valdez A-5
Approx. Area: 10 sq km (2,500 acres)

Directions: About 48 kilometers (30 mi) northeast of Valdez on Richardson Highway there is a turnoff to the Worthington Glacier

Recreation Area. A short trail from the parking lot will bring you to the glacier's side.

Notes: You can see deep into the blue-walled crevasses while standing on firm ground next to the glacier.

YAKUTAT AREA

Tongass National Forest
USDA Forest Service
Yakutat Ranger District
P.O. Box 327
Yakutat, AK 99869
(907) 784-3359

HUBBARD

Type: valley, tidewater
Lat/Long: 60° 15' / 138° 30'
Map: 1:63,360 USGS Mt. St Elias A-4
Approx. Area: 1,000 sq km (250,000 acres)

Directions: Boat into Disenchantment Bay or plane from Yakutat.

Notes: This is one of the longest tidewater glaciers in North America and it currently is advancing slowly. In 1986 the Hubbard's advancing front blocked Russell Fjord and trapped many sea animals for several months. The ice dam then broke through and drained the trapped water and animals. However, the Hubbard remains healthy with a 95 percent AAR, is still advancing, and could block the fjord again.

JUNEAU AREA

Tongass National Forest
USDA Forest Service
Juneau Ranger District
8465 Old Dairy Road
Juneau, AK 99801
(907) 789-3111

The Juneau Icefield covers approximately 1,800 square kilometers (450,000 acres). Although small by many other standards in Alaska, this area is one of the most well-studied in the United States. Several universities conduct regular research programs here and a couple of the glaciers have extended mass-balance studies.

Mendenhall Glacier Visitor Center

P.O. Box 2097
Juneau, AK 99803

MENDENHALL

Type: valley, calving
Lat/Long: 59° 30' / 134° 31'
Map: 1:63,360 USGS Juneau C-2 & B-2
Approx. Area: 100 sq km (25,000 acres)

Directions: The visitor center is 21 kilometers (13 mi) north of downtown Juneau and about 10 kilometers (6 mi) from the Alaska Marine Highway terminal. It can be reached by private vehicle, taxi service, or public transportation.

Notes: This glacier is as popular to visit as Canada's Athabasca Glacier. The calving terminus rises nearly 30 meters (100 ft) above Mendenhall Lake and can be seen from the visitor center. Several short trails bring you closer to the glacier. Although the glacier is retreating, its terminus remains steep because of calving into the lake. There are many splay crevasses that make access hazardous. The glacier is a tributary of the Juneau Icefield.

Wrangell and St. Elias

(see also Yukon Territory)

Wrangell St. Elias National Park and Preserve
P.O. Box 29
Glennallen, AK 99588
(907) 822-5234

BERING

Type: valley, piedmont
Lat/Long: 60° 20' / 143° 00'
Map: 1:250,000 USGS Bering Glacier
Approx. Area: 5,000 sq km (1.2 million acres)

Directions: This is a remote glacier that is best viewed from the air. It can usually be seen on commercial flights between Anchorage and Seattle, local air travel, or scenic flight services.

Notes: Many believe that this is the largest glacier in North America.

SEWARD–MALASPINA

Type: valley, piedmont
Lat/Long: 60° 00' / 141° 00'
Map: 1:250,000 USGS Mt. St Elias & Yakutat
Approx. Area: 5,000 sq km (1.2 million acres)

Directions: This also is a remote glacier that is best viewed from the air. As a neighbor of Bering, the Malaspina should be visible from the same flight pattern. A less spectacular view can be obtained on the ground from Yakutat.

Notes: This large piedmont glacier has enough surface area to cover the state of Delaware. Ice crystals found near its terminus have measured over 20 centimeters (8 in) in diameter.

KENNICOTT

Type: valley
Lat/Long: 61° 35' / 143° 15'
Map: 1:250,000 USGS McCarthy
Approx. Area: 300 sq km (75,000 acres)

Directions: Drive approximately 40 kilometers (25 mi) from Chitina to the Gilahina creek, then go another 50 kilometers (31 mi) over gravel road to the Kennicott River. The bridge was washed out some time ago so you will have to leave your car and use the hand-pull tramway to get across to the town of McCarthy. The Kennicott Glacier flows to the edge of town and should not be difficult to see. You can travel along the road to the town of Kennicott along the glacier's eastern flank to get a better view.

Notes: This is a large and very interesting glacier. Geysers have been observed.

California

Yosemite National Park

National Park Service
P.O. Box 577
Yosemite, CA 95389
(209) 372-0200

LYELL

Type: cirque
Lat/Long: 37° 44' / 119° 16'
Map: 1:24,000 USGS Mount Lyell
Approx. Size: 1 sq km (250 acres)

Directions: From Tuolumne Meadows follow the Lyell Canyon Trail. Just before Donoghue Pass, when you can see the glacier, cut cross-country to reach its terminus.

Notes: This is the largest glacier in Yosemite and is a remnant of the once great Tuolumne Glacier. It is believed that a longhorn sheep died and froze into the ice, only to melt out after it had become extinct in the park. Discovered in 1933 by two park naturalists, they surmised that it had been part of the glacier ice for fifty to two hundred and fifty years.

Mt. Shasta

District Ranger
Mt. Shasta Ranger District
204 West Alma St.
Mt. Shasta, CA 96067
(916) 926-4511

WHITNEY

Type: young valley
Lat/Long: 41° 25' / 122° 12'
Map: 1:24,000 USGS Mount Shasta
Approx. Area: 3 sq km (740 acres)

Directions: From U.S. Highway 97 turn south onto Military Pass Road. From Military Pass drive east on a logging road to the North Gate trailhead. Hike along the Bolan Wilderness Trail to above tree line then travel southwest cross-country to glacier terminus.

Notes: This is a very young glacier, having formed since Mt. Shasta's last major eruption about eight hundred years ago. Because it is so young, it has not yet carved a deep valley into the mountainside. The altitude and climate favor the growth of ablation hollows, and penitents up to shoulder depth have been observed.

Colorado

Boulder

City of Boulder
Department of Utilities
P.O. Box 791
Boulder, CO 80306
or

INSTAAR
Campus Box 450
University of Colorado
Boulder, CO 80309

ARAPAHOE

Type: cirque
Lat/Long: 40° 01' / 105° 38'
Map: 1:24,000 USGS Monarch Lake
Approx. Area: 0.5 sq km (125 acres)

Directions: From the junction of Colorado 72 and 119 at Nederland, drive north on 72 for 11.5 kilometers (7.2 mi). Turn left at University of Colorado Camp. Continue over a boulder-strewn road 8.3 kilometers (5.2 mi) to the trailhead at the far west end of Rainbow Lakes Campground. Follow "Glacier Rim/Arapaho Pass" trail to a saddle to view the glacier. The Arapahoe is in Boulder's watershed and therefore access is restricted.

Notes: Many people in this area are proud that the Arapahoe helps to supply Boulder's water. The Hotel Boulderado has a drinking fountain in its lobby that confidently states, "Pure cold water from Boulder owned Arapaho Glacier." In fact, the Arapahoe's meltwater supplies only a small fraction of the city's total water.

Arapahoe National Forest

USDA Forest Service
240 West Prospect Street
Fort Collins, CO 80526
(303) 224-1100

ST. MARY'S

Type: glacieret or perennial snow field
Lat/Long: 39° 50' / 105° 38'
Map: 1:24,000 USGS Empire
Approx. Area: less than 0.1 sq km (25 acres)

Directions: 3.2 kilometers (2 mi) east of Idaho Springs on I-70 take exit 238 and follow Fall River Road 14.4 kilometers (9 mi) to the St. Mary's ski area. Park in the ski area lot. The trail is about 100 meters (300 ft) north.

Notes: This patch of snow is a popular playground for summer skiing, snow boarding, and sledding. Because of its small size and apparent lack of movement, it may not be a true glacier.

Maroon Bells Wilderness

White River National Forest
Old Federal Building
Box 948
Glenwood Springs, CO 81602
(303) 945-2521

The Elk Mountains and Mt. Sopris are loaded with rock glaciers. You can see them from a distance by travelling south on Colorado 82 from Glenwood Springs to Aspen, but it may be worth hiking through the Maroon Bells Wilderness for close-up views.

Montana

Glacier National Park

National Park Service
West Glacier, MT 59936
(406) 888-5441

GRINNELL

Type: cirque
Lat/Long: 48° 45' / 113° 43'
Map: 1:24,000 USGS Logan Pass & Many Glacier
Approx. Area: 1 sq km (250 acres)

Directions: From U.S. Route 89, turn west at Babb. Drive 20 kilometers (12 mi) to the park's Many Glacier area. From there you can follow the Grinnell Glacier Trail or join a Ranger for a guided trip that begins by boat.

Notes: Although small, this glacier is nestled under the steep rock cliffs of the Continental Divide. It is nourished by wind-transported snow from prevailing westerlies and by avalanches from the surrounding slopes.

Beartooth Range

Gallatin National Forest
Federal Building
Box 130
Bozeman, MT 59771
(406) 587-6701

GRASSHOPPER

Type: cirque
Lat/Long: 45° 07' / 109° 52'
Map: 1:24,000 USGS Little Park Mountain
Approx. Area: 0.5 sq km (125 acres)

Directions: From U.S. Highway 12, about 3 kilometers (2 mi) east of Cooke City, turn north at the Cooke Ranger Station. Within about 3 kilometers (2 mi) this gravel road will take you to an unmaintained jeep trail, which you should follow to Goose Lake, approximately 8 kilometers (5 mi). A trail from Goose Lake approaches the glacier from a saddle just north of Iceberg Peak.

Notes: This glacier, and several others on the Beartooth Plateau, are so named for the embedded grasshoppers and grasshopper parts. It appears that summer migration of grasshoppers from the nearby plains coincides with an annual cycle of heavy thunderstorms in August. The grasshoppers crash into the ice and are frozen into the glacier.

NEVADA

GREAT BASIN NATIONAL PARK

Baker, NV 89311
(702) 234-7331

WHEELER

Type: cirque
Lat/Long: 38° 59' / 114° 18'
Map: 1:24,000 USGS Wheeler Peak
Approx. Area: 0.2 sq km (50 acres)

Directions: From U.S. Highway 50/6 turn south onto State Route 487 to Baker, then follow signs to the park, about 8 kilometers (5 mi). During summer, drive to Wheeler Peak Campground then follow trail for about 5 kilometers (3 mi) to the glacier. During winter, ski from Lehman Campground about 6 kilometers (4 mi) to the Wheeler Peak Campground.

Notes: This is considered the last remaining glacier in the Great Basin. There is great controversy as to whether it is truly a glacier, mainly because it no longer appears to move. There are crevasses, but they are old. The west section is most certainly stagnant.

OREGON

MT. HOOD NATIONAL FOREST

Zig-Zag Ranger District
Rhododendrum, OR 97049
(503) 666-0704

There are nine major glaciers on Mt. Hood. All are relatively easy to access via well-maintained hiking trails around the mountain.

PALMER

Type: valley
Lat/Long: 45° 21' / 121° 42'
Map: 1:24,000 USGS Mt. Hood South
Approx. Area: 1.5 sq km (370 acres)

Directions: Follow U.S. Highway 26 to Government Camp then turn north to Timberline Lodge. Ski lifts access the glacier during summer.

Notes: Timberline is oldest year-round ski resort in North America and affords that luxury because of Mt. Hood's glaciers. Since most ski areas face north to help extend their seasons, it is interesting that Timberline faces south. Actually, skiing is on a remnant lobe of the Palmer Glacier that may now only be a perennial patch of snow.

WASHINGTON

MT. RAINIER NATIONAL PARK

Longmire, WA 98397
(206) 569-2211

CARBON

Type: valley
Lat/Long: 46° 53' / 121° 46'
Map: 1:24,000 USGS Mowich Lake
Approx. Area: 10 sq km (2,500 acres)

Directions: Follow State Route 410 from Tacoma east to Buckley. Turn south on State Route 165 and follow signs to Mt. Rainier National Park's Carbon River Entrance (closed during winter). From the parking lot, the Carbon River Trail takes you within view of the glacier in about 4 kilometers (2.5 mi). A larger portion of the glacier can be seen by following the trail toward Moraine Park.

Notes: The Carbon is aptly named because of its blackened, dirt-covered surface. This, along with its steep valley and northerly aspect, make it one of Washington's healthiest glaciers. Small thrust faults are visible on its terminus face. Its ominously dark advancing front can be seen from a distance towering through the trees.

INGRAHAM/COWLITZ

Type: valley
Lat/Long: 46° 50' / 121° 43'
Map: 1:24,000 USGS Mt. Rainier East
Approx. Area: 12 sq km (3,000 acres)

Directions: Follow State Route 7 from Tacoma southeast to Elbe, then travel east on State Route 706 to Mt. Rainier's Longmire entrance (open year-round). Drive to the Paradise Visitor Center. From there follow a short trail to the Paradise Ice Caves then traverse across the Stevens Glacier to a saddle just north of the Cowlitz Rocks for a view down onto the Ingraham and Cowlitz Glaciers.

Notes: The ogive pattern that can be seen from Cowlitz Rocks is the most easily visible in North America. Although they are considered part of the Cowlitz Glacier, the ogives are actually formed from Ingraham Glacier's icefall and separated from the main flow of the Cowlitz by a medial moraine. Spring skiing is popular on the Cowlitz névé from Camp Muir.

NISQUALLY

Type: valley
Lat/Long: 46° 50' / 121° 45'
Map: 1:24,000 USGS Mt. Rainier East
Approx. Area: 6 sq km (1,500 acres)

Directions: Follow State Route 7 from Tacoma southeast to Elbe, then travel east on State Route 706 to Mt. Rainier's Longmire entrance (open year-round). Drive to the Paradise Visitor Center. Full view of the glacier is available from several places along the road toward Paradise. Several short trails from the visitor center take you closer.

Notes: This is a steep valley glacier so it can be seen in its entirety, from tip to toe. Because it is part of the Tacoma watershed, its fluctuations are monitored regularly. Ice from its snout used to supply the Longmire Inn.

PARADISE/STEVENS

Type: stagnant remnant, ice caves
Lat/Long: 46° 48' / 121° 42'
Map: 1:24,000 USGS Mt. Rainier East
Approx. Area: 1.5 sq km (370 acres)

Directions: Follow State Route 7 from Tacoma southeast to Elbe, then travel east on State Route 706 to Mt. Rainier's Longmire entrance (open year-round). Drive to the Paradise Visitor Center. From there follow a short trail to the Paradise Ice Caves.

Notes: Although called the Paradise Ice Caves, these caves are actually in a remnant portion of the Paradise Glacier called the Stevens Glacier or the Stevens Lobe. The caves are constantly changing and can be unstable. They offer a fascinating view of the glacier's underbelly.

Mt. St. Helens Monument

Visitor Center
3029 Spirit Lake Highway
Castle Rock, WA 98611
(206) 274-6644

There are nine major glaciers on Mt. St. Helens. Nearly all lost their accumulation zones during the 1980 eruption. Several became dirt-covered and are now protected from further ablation. New glaciers are forming inside the crater wall.

SHOESTRING

Type: debris-covered valley
Lat/Long: 46° 11' / 121° 8'
Map: 1:24,000 USGS Mt. St. Helens
Approx. Area: 1 sq km (250 acres)

Directions: From Interstate 5, drive east on U.S. Highway 12 to Randle. Turn south onto Forest Service Road 25 for approximately 13 kilometers (8 mi) to Forest Service Road 26. Follow 26 to the end at Windy Pass, about 30 kilometers (20 mi). Hike south on Trail 216 for about 6 kilometers (4 mi). This will get you to the toe of Shoestring.

Notes: The accumulation zone of this glacier was blown all over eastern Washington during Mt. St. Helens' 1980 eruption. The rest of it was covered by ash and pumice. The volcanic debris covering has helped the glacier survive subsequent ablation seasons.

Wyoming

Wind River Range

Fitzpatrick Wilderness
Wind River Ranger District
Box 186
Dubois, WY 82513
(307) 455-2466

DINWOODY

Type: valley
Lat/Long: 43° 10' / 109° 38'
Map: 1:24,000 USGS Gannett Peak
Approx. Area: 4 sq km (1,000 acres)

Directions: From U.S. Highway 287/26 drive about 26 kilometers (16 mi) to the Trail Lake trailhead. Follow the Glacier Trail toward Gannett Peak about 24 kilometers (15 mi).

Notes: The central portion of Dinwoody is nearly flat with "snow swamps." Because this is a small glacier with a simple stress pattern, there are hundreds of annual layers visible in the ablation zone. There are also a couple of medial moraines.

HEAP STEEP

Type: cirque
Lat/Long: 43° 10' / 109° 37'
Map: 1:24,000 USGS Fremont Peak North
Approx. Area: 0.1 sq km (25 acres)

Directions: From U.S. Highway 287/26 drive about 26 kilometers (16 mi) to the Trail Lake trailhead. Follow the Glacier Trail toward Gannett Peak about 20 kilometers (12 mi).

Notes: Like the Dinwoody, the small Heap Steep has many annual layers that are visible near its terminus.

Glossary

Ablation. Any process by which a glacier loses mass including melt, evaporation, sublimation, and calving. Often called wastage.

Ablation area. The area of a glacier where more mass is lost than gained.

Ablation hollows. Depressions in the snow surface caused by sun or warm, gusty wind. *See* sun cups and penitents.

Ablation moraine. A mound or layer of moraine in the ablation zone of a glacier. The rock has been plucked from the mountainside by the moving glacier and is melting out on the ice surface.

Ablation season. A period during which glaciers lose more mass than gain; usually coincides with summer.

Ablation zone. *See* ablation area.

Accumulation. Any process by which a glacier gains mass. Usually in the form of snowfall but also by avalanching, wind-drifted snow, tributary glaciers, etc.

Accumulation area. The area of a glacier where more mass is gained than lost. Also called the névé of a glacier.

Accumulation area ratio (AAR). The ratio between the area of accumulation and the total area of the glacier. Used as a measure of glacier health.

Accumulation season. A period during which a glacier gains more mass than it loses; usually coincides with winter.

Accumulation zone. *See* accumulation area.

Active rope. The side of a protective climbing rope that is attached to the moving partner in a belay.

Advance. *See* glacial advance.

Alpine glacier. *See* mountain glacier.

Annual layers. Annual accumulations of snow and dust on a glacier.

Band ogives. Alternate bands of light and dark on a glacier. Usually found below steep narrow icefalls and thought to be the result of different flow and ablation rates between summer and winter. Also called Forbes bands.

Basal plane. In an ice crystal molecules are concentrated close to a series of parallel planes called basal planes.

Bergschrund. A crevasse that separates flowing ice from stagnant ice at the head of a glacier.

Bergy bit. A rounded iceberg that is about the same size as a two-room cabin and floats with less than 5 meters (15 ft) showing above sea level.

Bottom bergs. Icebergs that originate from near the bottom of a glacier. The color is usually black from trapped rock material or dark blue because of old, coarse, bubble-free ice. They sit low in the water due to the weight of the embedded rocks.

Branched-valley glacier. A glacier that has one or more tributary glaciers that flow into it. Distinguished from a simple valley glacier.

Calving. A process by which ice breaks off a glacier's terminus. Usually the term is reserved for tidewater glaciers or glaciers that end in lakes, but it can refer to the serac falls from hanging glaciers.

Calving glacier. A glacier that loses material by calving. Usually a glacier that terminates in sea, lake, or river water.

Catchment glacier. A glacier that receives nourishment from wind-blown snow.

Cirque. A bowl shape or amphitheater usually sculpted out of the mountain terrain by a cirque glacier.

Cirque glacier. A glacier that resides in basins or amphitheaters near ridge crests. Most have a characteristic circular shape, with its width as wide or wider than its length.

Compressing flow. When glacier motion is decelerating down-slope.

Constructive metamorphism. Snow metamorphism that adds molecules to sharpen the corners and edges of an ice crystal. *See* kinetic-growth metamorphism.

Creep. This is a way that snow or ice can move by deforming its internal structure.

Crevasse. An open fissure in the glacier surface. *See also* marginal crevasse, splay crevasse, and transverse crevasse.

Depth hoar. An ice crystal that develops within a layer of snow. It is characterized by rapid recrystallization, usually caused by strong gradients in temperature, forming crystal shapes that resemble cups and scrolls. Typically found near the bottom of an annual accumulation of snow and most persistent on polar or subpolar glaciers where air temperatures are cold and annual snow accumulations are light.

Destructive metamorphism. Snow metamorphism that rounds the corners and edges of an ice crystal. *See* radius-dependent metamorphism.

Dirt cone. A cone-shaped formation of ice that is covered by dirt. It is caused by a differential pattern of ablation between the dirt-covered surface and bare ice.

Drain channel. 1. A preferred path for meltwater to flow from the surface through a snow cover. 2. *See also* meltwater conduit.

Drift glacier. *See* catchment glacier.

Dump moraine. A mound or layer of moraine formed along the edge of a glacier by rock that falls off the ice. Sometimes called a ground moraine.

End moraine. An arch-shaped ridge of moraine found near the end of a glacier.

Equilibrium line. A boundary between the accumulation area and ablation area where the mass balance is zero.

Equilibrium metamorphism. *See* radius-dependent metamorphism.

Equi-temperature metamorphism. Snow metamorphism that occurs under relatively consistent temperature conditions. *See* radius-dependent metamorphism.

Extending flow. When glacier motion is accelerating down-slope.

False ogives. Bands of light and dark on a glacier that were formed by rock avalanching.

Firn. Rounded, well-bonded snow that is older than one year. It has a density greater than 550 kg/m^3 (35 lb/ft^3). Also called névé.

Firn limit. *See* firn line.

Firn line. The minimum elevation of firn lying on a glacier surface. Each year's firn line marks a glacier's annual equilibrium line.

Firn water table. The height of meltwater within saturated firn that is trapped over ice in a glacier.

Firnspiegel. A thin sheet of ice formed on the glacier surface by rapid refreezing of solar-heated snow or firn, usually at high elevations during spring.

Flow finger. A small percolation channel that is a beginning path for surface meltwater through snow or firn. *See also* drain channel.

Fluted berg. An iceberg that is grooved into a curtain-like pattern. Thought to be carved by small meltwater streams.

Foliation. Layering in glacier ice that has distinctive crystal sizes and/or number of bubbles. Usually caused by stress and deformation that a glacier experiences as it flows over complex terrain, but can also originate as a sedimentary feature.

Forbes bands. *See* band ogives.

Forel stripes. Shallow, parallel grooves on the face of a large melting ice crystal.

Free water. Liquid water in or around ice. It is usually from the melting ice.

Geyser. A fountain that develops when water from a conduit is forced up to the surface of a glacier. Also called a negative mill.

Glacial advance. When a mountain glacier's terminus extends farther down-valley than before. This occurs when a glacier flows down-valley faster than the rate of ablation at its terminus.

Glacial retreat. When the position of a mountain glacier's terminus is farther up-valley than before. This occurs when a glacier ablates more material at its terminus than it transports into that region.

Glacial till. Accumulations of unsorted, unstratified mixtures of clay, silt, sand, gravel, and boulders. The usual composition of moraines.

Glacier. A mass of ice that originates on land. Usually having an area larger than one tenth of a square kilometer. Many believe that a glacier must show some type of movement. Others believe that a glacier can show evidence of past or present movement. *See also* alpine glacier, branched-valley glacier, catchment glacier, cirque glacier, drift glacier, hanging glacier, mountain glacier, niche glacier, piedmont glacier, polar glacier, rock glacier, subpolar glacier, surging glacier, temperate glacier, tidewater glacier, and valley glacier.

Glacier cave. A cave of ice. Usually underneath a glacier and formed by meltwater. Cave entrances are often enlarged near a glacier terminus by warm winds. Most common on stagnant portions of glaciers.

Glacier fire. A phenomenon in which strong reflection of the sun on an icy surface causes a glacier to look like it is on fire.

Glacier flour. A fine powder of silt- and clay-sized particles that a glacier creates as its rock-laden ice scrapes over bedrock. It is usually flushed out in meltwater streams and causes water to look powdery gray. Lakes and oceans that fill with glacier flour may develop a banded appearance. Also called rock flour.

Glacier ice. Well-bonded ice crystals compacted from snow with a bulk density greater than 860 kg/m^3 (55 lb/ft^3). Air becomes trapped inside the crystal fabric in the form of bubbles.

Glacier mill. *See* moulin.

Glacier remainié. A glacier that is reconstructed or reconstituted out of other glacier material. Usually formed by seracs falling from a hanging glacier then re-adhering.

Glacier snout. *See* glacier terminus.

Glacier sole. The bottom of the ice of a glacier.

Glacier table. A rock that resides on a pedestal of ice. Formed by differential ablation between the rock-covered ice and surrounding bare ice.

Glacier terminus. The lowest end of a glacier. Also called the glacier toe or glacier snout.

Glacier toe. *See* glacier terminus.

Glacier wind. 1. *See* katabatic wind. 2. Wind that flows out of ice caves.

Glacieret. A very small glacier.

Ground moraine. A continuous layer of till near the edge or underneath a steadily retreating glacier.

Growler. An iceberg less than 2 meters (7 ft) across that floats with less than 1 meter (3 ft) showing above water. Smaller than a bergy bit.

Hanging glacier. A glacier that terminates at or near the top of a cliff.

Headwall. A steep cliff, usually the uppermost part of a cirque.

Ice. The solid crystalline form of water.

Ice apron. A mass of ice adhering to a mountainside.

Ice cap. A dome-shaped mass of glacier ice that spreads out in all directions. It is usually larger than an icefield but less than 50,000 square kilometers (12 million acres).

Ice cave. *See* glacier cave.

Ice quake. A shaking of ice caused by crevasse formation or jerky motion.

Ice sheet. A dome-shaped mass of glacier ice that covers surrounding terrain and is greater than 50,000 square kilometers (12 million acres) (e.g., Greenland and Antarctic ice sheets).

Ice shelf. That portion of an ice sheet that spreads out over water.

Ice stream. 1. A current of ice in an ice sheet or ice cap that flows faster than the surrounding ice. 2. Sometimes referring to the confluent sections of a branched-valley glacier. 3. Obsolete synonym of valley glaciers.

Ice worm. An oligochaete worm that lives on temperate glaciers or perennial snow. There are several species that range in color from yellowish brown to reddish brown or black. They are usually less than 1 millimeter (.04 in) in diameter and average about 3 millimeters (0.1 in) long. Some feed off red algae.

Iceberg. A piece of ice that has broken off from the end of a glacier that terminates in water.

Ice-cemented glacier. *See* rock glacier.

Ice-cored glacier. *See* rock glacier.

Icefall. A steep, fast-flowing section of glacier that usually has a cracked and jumbled surface.

Icefield. A mass of glacier ice that flows outward in all directions. Distinguished from an ice cap in that it is usually smaller, somewhat controlled by terrain, and does not have a dome-like shape.

Inactive rope. The side of protective climbing rope that is tied to the belayer.

Jökulhlaup. 1. A large outburst flood that usually occurs when a glacially dammed lake drains catastrophically. 2. Any catastrophic release of water from a glacier. *See* outburst flood.

Katabatic wind. Wind that flows from a glacier. It is caused by air that cools over the ice surface becoming heavier than surrounding air, then draining down-valley. Also called glacier wind.

Kinetic-growth metamorphism. Snow metamorphism that builds angular facets on crystals and makes cup and scroll shaped crystals. *See* temperature-gradient metamorphism.

Lateral moraine. A ridge-shaped moraine, deposited at the side of a glacier, composed of material that was eroded from the valley walls by the moving glacier.

Marginal crevasse. A crevasse near the side of a glacier formed as the glacier moves past stationary valley walls. Usually oriented about 45 degrees up-glacier from the side wall.

Mass balance. The difference between accumulation and ablation on a glacier; usually calculated on an annual basis.

Medial moraine. A ridge-shaped moraine in the middle of a glacier originating from a rock outcrop, nunatak, or the converging lateral moraines of two or more ice streams.

Meltwater conduit. A channel within, underneath, on top of, or near the side of a glacier that drains meltwater out of the glacier. It is usually kept open by the frictional heating of flowing water that melts the ice walls of the conduit.

Moat. A deep trench or channel at the edge of a shrinking glacier.

Moraine. A mound, ridge, or other distinct accumulation of glacial till. *See also* ablation moraine, ground moraine, lateral moraine, medial moraine, and push moraine.

Moraine shoal. A glacial moraine that has formed a shallow place in water.

Moulin. A nearly vertical channel in ice that is formed by flowing water. Usually found after a relatively flat section of glacier in a region of transverse crevasses. Also called a glacier mill.

Mountain glacier. A glacier that is confined by surrounding mountain terrain.

Negative mill. *See* geyser.

Névé. 1. The accumulation zone of a glacier. 2. Firn.

Niche glacier. A glacier that resides in a small recess of the terrain. Also called a pocket glacier.

Ogives. Alternate bands of light and dark ice seen on a glacier surface. *See* band ogives, false ogives, sedimentary ogives, and wave ogives.

Outburst flood. Any catastrophic flooding from a glacier. May originate from trapped water in cavities inside a glacier or at the margins of glaciers or from lakes that are dammed by flowing glaciers. *See also* jökulhlaup.

Penitents. The extreme relief of ablation hollows found most often at high altitudes in the tropics. The resulting spikes of snow resemble repentant souls.

Perennial snow. Snow that persists on the ground year after year.

Piedmont glacier. A large ice lobe that is spread out over surrounding terrain, associated with the terminus of a large mountain valley glacier.

Pocket glacier. *See* niche glacier.

Polar glacier. A glacier whose temperatures are below freezing throughout, except possibly for a thin layer of melt near the surface during summer or near the bed. Found only in polar regions of the globe or at high altitudes.

Pothole. *See* moulin.

Pressure melting. Melting that occurs in ice at temperatures colder than normal melting temperature because of added pressure.

Push moraine. A moraine built out ahead of an advancing glacier.

Radius-dependent metamorphism. Snow metamorphism that occurs when there are large differences in convex and concave portions of a crystal.

Randkluft. A fissure that separates a moving glacier from its headwall rock. Like a bergschrund.

Reconstituted glacier. *See* glacier remanié.

Reconstructed glacier. *See* glacier remanié.

Red algae. An algae that is common on temperate glaciers and perennial snow. Its red color sometimes prompts people to call it "watermelon snow."

Regelation. Motion of an object through ice by melting and freezing that is caused by pressure differences. This process allows a glacier to slide past small obstacles on its bed.

Regenerated glacier. *See* glacier remanié.

Retreat. *See* glacial retreat.

Rock flour. *See* glacier flour.

Rock glacier. A glacier whose motion and behavior is characterized by a large amount of embedded or overlying rock material. 1. Ice-cemented rock formed in talus that is subject to permafrost. 2. Ice-cemented rock debris formed from avalanching snow and rock. 3. Rock debris that has a core of ice; either a debris-covered glacier or a remnant end moraine.

Seasonal snow. 1. Snow that accumulates during one season. 2. Snow that lasts for only one season.

Sedimentary ogives. Alternating bands of light and dark at the firn limit of a glacier. The light bands are usually young and lightest at the highest level up-glacier, becoming increasingly older and darker as they progress down-glacier.

Self-arrest. A technique that uses an ice axe to help a person stop sliding on snow or ice.

Serac. An isolated block of ice that is formed where the glacier surface is fractured.

Sintering. The bonding together of ice crystals.

Slush. A mixture of about half snow and half water.

Slush zone. Common near the snow line on a relatively flat portion of a glacier where melting snow forms slush.

Snow. 1. An ice particle formed by sublimation of vapor in the atmosphere. 2. A collection of loosely bonded ice crystals deposited from the atmosphere. High density snow (greater than 550 kg/m^3 [35 lb/ft^3]) is called firn if it is older than one year.

Snow layer. A layer of ice crystals with similar size and shape.

Snow line. The minimum elevation of snow lying on the ground or glacier surface. The snow line at the end of an ablation season marks a glacier's current equilibrium line.

Snow worm. *See* ice worm.

Spelunking. The activity of exploring caves.

Splay crevasse. A crevasse pattern that forms where ice slowly spreads out sideways. Commonly found near the terminus of glaciers.

Sublimation. The change of state from ice to water vapor or water vapor to ice.

Subpolar glacier. A glacier whose temperature regime is between polar and temperate. Usually predominantly below freezing, but could experience extensive summer melt.

Sun cups. Ablation hollows that develop during intense sunshine.

Surging glacier. A glacier that experiences a dramatic increase in flow rate, ten to one hundred times faster than its normal rate. Usually surge events last less than one year and occur periodically, between fifteen and one hundred years.

Talus. Coarse, angular rock fragments that are loose. Usually from a cliff or very steep, rocky slope and therefore often found at its base.

Talus glacier. *See* rock glacier.

Temperate glacier. A glacier that is near its freezing temperature throughout, except for a thin layer of below-freezing temperatures near its surface during winter. Typically experiences extensive melt and water drainage through a complex plumbing system.

Temperature-gradient metamorphism. Snow metamorphism that occurs when there are strong differences in temperature between the bottom and top of a snow layer. *See* kinetic-growth metamorphism.

Thomson crystal. A large ice crystal found in deep, stagnant water-filled cavities of a glacier.

Tidewater glacier. A mountain glacier that terminates in the ocean.

Till. *See* glacial till.

Transverse crevasse. A crevasse that forms across a glacier in a region where velocity increases down the glacier. Common in the accumulation zone and near steepening slopes.

Tyndall figures. Small melt features that develop within an ice crystal.

Tyndall flowers. Melt features within an ice crystal that develop branched shapes, usually with hexagonal symmetry.

Valley glacier. A mountain glacier whose flow is confined by valley walls.

Wastage. *See* ablation.

Water table. *See* firn water table.

Wave ogives. Ogives that show some vertical relief on a glacier. Usually the dark bands are in the hollows and the light bands are in the ridges. Formed at the base of steep, narrow icefalls.

Weathered ice. Glacier ice that has been exposed to sun or warm wind so that the boundaries between its ice crystals are partly disintegrated.

Selected Bibliography

Glaciers

Bishop, G., and J. Forsyth. 1988. *Vanishing Ice*. Dunedin, New Zealand: John McIndoe Ltd. in association with New Zealand Geological Survey. 56 pp.

This introductory book is based upon a study of New Zealand's Dart Glacier. It is well written and informative.

Giardino, J. R., and others. 1987. *Rock Glaciers*. London: Allen and Unwin. 355 pp.

This is a college-level text on rock glaciers and discusses some of the bases for controversy.

LaChapelle, E. R., and A. Post. 1971. *Glacier Ice*. Seattle, Wash.: The Mountaineers. 110 pp.

Unfortunately out of print, this pictorial book has the right combination of beautiful pictures and a well-written, explanatory text. Although there are rumors of a reprint, I would jet off to the nearest library for a look as soon as possible.

Paterson, W.S.B. 1981. *The Physics of Glaciers*, 2d ed. New York: Pergamon Press. 380 pp.

This is a standard text about glaciers and ice sheets. It is for those contemplating research in the subject so it assumes basic knowledge of physics.

Sharp, R. P. 1960. *Glaciers*. Eugene, Ore.: University of Oregon Books. 78 pp.

This is a classic. Although technical principles of glaciers are described, less knowledge of physics is assumed.

Sharp, R. P. 1988. *Understanding Glaciers and Glaciation.* New York: Cambridge University Press. 225 pp.

As an extension of Sharp's earlier book this one includes many of the same photos. There is also an expanded section on glacial geology.

Guide Books

Alaska Geographic. 1982. *Alaska's Glaciers.* Vol. 9, no. 1. Anchorage: Alaska Geographic Society. 114 pp.

Bailey, E., and K. VanTighem. 1987. *Columbia Icefield—Ice Apex of the Canadian Rockies.* Jasper: Parks and People in association with Environment Canada. 48 pp.

Conner, C., and D. O'Hare. 1988. *Roadside Geology of Alaska.* Missoula, Mont.: Mountain Publishing Company. 250 pp.

Driedger, C. L. 1986. *A Visitor's Guide to Mount Rainier Glaciers.* Longmire, Wash.: Pacific Northwest National Parks and Forests Association. 80 pp.

Gadd, Ben. 1986. *Handbook of the Canadian Rockies.* Jasper, Alberta: Corax Press. 876 pp.

Kirk, Ruth. 1983. "Of Time and Ice," in *Glacier Bay: A Guide to Glacier Bay National Park and Preserve, Alaska.* Handbook 123. Washington, D.C.: Division of Publications, National Park Service. 128 pp.

Kucera, R. E. 1981. *Exploring the Columbia Icefield.* Canmore, Alberta: High Country Publishers. 64 pp.

Lethcoe, N. 1987. *An Observers Guide to Glaciers of Prince William Sound, Alaska.* Valdez, Alaska: Prince William Sound Books. 151 pp.

Weather and Climate

Barry, R. G. 1981. *Mountain Weather and Climate.* New York: Methuen. 313 pp.

Imbrie, J., and K. Imbrie. 1986. *Ice Ages: Solving the Mystery.* Cambridge: Harvard University Press. 224 pp.

Keen, R. 1989. *Skywatch: A Guide to Western Weather.* Golden, Colo.: Fulcrum, Inc.

Reifsnyder, W. E. 1980. *Weathering the Wilderness: The Sierra Club Guide to Practical Meteorology.* San Francisco: Sierra Club Books. 276 pp.

Emergency Medicine

Auerbach, Paul S., M.D. 1986. *Medicine for the Outdoors.* Boston: Little, Brown and Co. 347 pp.

Hackett, Peter H., M.D. 1984. *Mountain Sickness.* New York: American Alpine Club. 75 pp.
Wilkerson, James A., M.D., ed. 1985. *Third Edition Medicine for Mountaineering.* Seattle, Wash.: The Mountaineers Books. 438 pp.

Mountaineering and Glacier Travel

Selters, Andy. 1990. *Glacier Travel and Crevasse Rescue.* Seattle, Wash.: The Mountaineers Books. 160 pp.
Williamson, J. E., and J. Whitteker, eds. 1989. *Accidents in North American Mountaineering.* New York and Banff: The American Alpine Club and the Alpine Club of Canada. 93 pp.

Other Resources

To find a competent guide service or training program, contact the Association of Canadian Mountain Guides or the American Mountain Guides Association. Their memberships are growing and both can refer you to the right place.

Association of Canadian Mountain Guides
Box 1537
Banff, AB TOL OCO

American Mountain Guides Association
P.O. Box 2128
Estes Park, CO 80517

Subject Index

Geographic Index